BATTERY TECHNOLOGY

BATTERY TECHNOLOGY

From Fundamentals to Thermal Behavior and Management

MARC A. ROSEN
University of Ontario Institute of Technology, Oshawa, ON, Canada

AIDA FARSI
Faculty of Engineering and Applied Science, University of Ontario Institute of Technology, Oshawa, ON, Canada

ELSEVIER

ACADEMIC PRESS
An imprint of Elsevier

Academic Press is an imprint of Elsevier
125 London Wall, London EC2Y 5AS, United Kingdom
525 B Street, Suite 1650, San Diego, CA 92101, United States
50 Hampshire Street, 5th Floor, Cambridge, MA 02139, United States
The Boulevard, Langford Lane, Kidlington, Oxford OX5 1GB, United Kingdom

ISBN 978-0-443-18862-6

For information on all Academic Press publications
visit our website at https://www.elsevier.com/books-and-journals

Publisher: Charlotte Cockle
Acquisitions Editor: Edward Payne
Editorial Project Manager: Joshua Mearns
Production Project Manager: Fizza Fathima
Cover Designer: Miles Hitchen

Typeset by STRAIVE, India

To our families, friends, and colleagues, for their support and inspiration.

Contents

About the authors

Marc A. Rosen, PhD, is Professor at Ontario Tech University (officially University of Ontario Institute of Technology) in Oshawa, Ontario, Canada, where he served as founding Dean of the Faculty of Engineering and Applied Science. With more than 70 research grants and contracts and 900 technical publications, Dr. Rosen is an active teacher and researcher in sustainable energy, energy storage, thermodynamics, and environmental impact. Much of his research has been carried out for industry. Dr. Rosen has served as President of the Engineering Institute of Canada and of the Canadian Society for Mechanical Engineering. He has acted in many professional capacities, including Editor-in-Chief of various journals and Director of Oshawa Power and Utilities Corporation. Dr. Rosen has worked for such organizations as Imatra Power Company in Finland, Argonne National Laboratory near Chicago, the Institute for Hydrogen Systems near Toronto, and Toronto Metropolitan University (formerly Ryerson University) in Toronto, where he served as Chair of the Department of Mechanical, Aerospace, and Industrial Engineering. Dr. Rosen has received numerous awards and honors and is Fellow of numerous societies, including the Engineering Institute of Canada, Canadian Academy of Engineering, Canadian Society for Mechanical Engineering, American Society of Mechanical Engineers, International Energy Foundation, and Canadian Society for Senior Engineers.

Aida Farsi, PhD, is a postdoctoral fellow at Ontario Tech University, Oshawa, Canada, in the Faculty of Engineering and Applied Science. Her current research focuses on developing environmentally benign and sustainable systems for hydrogen production and storage. She also pursued her PhD degree in the area of clean hydrogen production via thermochemical copper chlorine cycle in which she developed an electrochemical model for the CuCl/HCl electrolyzer and a multiphase flow model for efficient recycling of the oxidized anolyte in the electrolyzer. With over 25 technical publications, Dr. Farsi is an active researcher in sustainable energy storage technologies and freshwater production systems.

Preface

Over the past few decades, with rising concerns regarding energy and the environment as well as advancements in cross-disciplinary communications between electrical and mechanical engineers and material scientists and electrochemists, the battery technology market has expanded greatly. It has also expanded from small- to large-scale applications during this period. At the same time, increasing greenhouse gas emissions from fossil fuel–based energy systems and concerns over energy security have led to the development of renewable energy–based systems for electricity production as well as the development of smart cities that use stable smart grids with clean sources of energy. Increasing the use of renewable energy–based systems usually leads to an increase in the use of energy storage systems, including battery systems, to balance the intermittent and fluctuating electrical power production and, consequently, to improve the reliability and overall usefulness of the energy systems. In addition, concerns over unstable energy prices and increasing carbon emissions from fossil fuel–based automobiles have led to significant growth in the utilization of electric vehicles and research in this area. The electric vehicle, including battery storage, is considered a highly promising technology for the future, with significant benefits to society, owning to its high efficiency, low emissions, and low noise.

Important needs for more widespread adoption of battery technology in automobiles and other applications include lowering the energy storage cost and environmental impacts, increasing the specific energy and energy density, enhancing power grid compatibility, improving fast charging, and ensuring safe operation. Achieving these needs requires a deep understanding of various types of battery cells and an ability to analyze them. It also requires an ability to optimize the design of battery packs to delay aging mechanisms in battery cells, to reduce or minimize uneven current distributions, to properly control discharge and charge operations, and to avoid risks of failures in battery packs.

In line with the aforementioned needs, this book describes battery technologies, with a focus on thermal behavior and sustainable approaches for efficient electrical energy storage via battery systems. Given that a significant interest in the thermal behavior of battery technologies and their thermal management is relatively new, it is particularly important that these areas receive increased attention. This is particularly important for battery applications in electric vehicles.

Battery Technology: From Fundamentals to Thermal Behavior and Management constitutes a comprehensive reference on rechargeable battery technology that describes important aspects of battery operation, analysis, performance, and application, including relevant aging mechanisms and thermal management systems. To provide a more holistic understanding and appreciation, thermodynamic, electrochemical, and thermal concepts and principles, which are important for the design, analysis, and optimization of battery systems, are considered throughout. Recent developments and advances in battery thermal management systems as well as the main design steps for a battery pack are identified and described. Finally, the book addresses the main problems regarding the impact of the thermal behavior of batteries on their life cycles and performance as well as challenges and future directions for sustainable development of battery technologies. Case studies and examples are incorporated throughout the book to help the reader understand and perform practical applications of battery technologies. As a consequence, this book provides a unique text and resource containing organized and synthesized knowledge about battery systems and their analysis that can guide students, researchers, and engineers in understanding and applying key concepts and cutting-edge developments in the field.

As outlined here, this book is structured so as to build a coherent understanding for the reader while progressing through it. In Chapter 1, the underlying fundamentals are described to provide an understanding of the development and operation of rechargeable battery systems. Also, general working principles of rechargeable battery systems as well as characteristics and applications of various battery types are covered. Chapter 2 is dedicated to the thermodynamics of battery technology. It includes definitions of enthalpy of reaction and Gibbs function of reaction for a battery system, derivation of the Nernst equation based on the first and second laws of thermodynamics for a generalized battery system, the development of a general energy balance for a battery system accounting for losses in the electrochemical reactions, and determination of the heat generation rate within a battery. In Chapter 3, an electrochemical model of a battery is developed, which provides a comprehensive description about the irreversibilities caused by overpotentials in a battery system, the heat generation rate in a battery system due to electrochemical reactions as well as ion and charge transfer processes, and the energy efficiency for a battery system. In Chapter 4, the thermal behavior of the battery is described, including the connection between the thermal behavior of a battery system and heat generation, aging mechanisms, thermal failure, and thermal management systems. In addition, a thermal-electrochemical coupled model of battery cells is developed, which allows for determination of the average cell temperature and the temperature distribution inside a battery cell. The primary degradation mechanisms of battery cells are described and discussed, as are challenges and opportunities regarding the impact of thermal issues on the performance and lifetime of batteries. Chapter 5 introduces conventional and modern battery thermal management technologies. Descriptions and comparisons are presented of conventional battery thermal management systems, such as air, liquid, and phase change material-based unit, as well as recently developed technologies, such as evaporative pool boiling-based battery thermal management systems. In Chapter 6, the main process steps in the design of a battery pack, including the design of the battery management system as well as electrical, mechanical, and thermal designs for a battery pack, are explained. The optimum design is discussed of a battery pack that can extend the life of the battery cells and provide safe and reliable operation of the battery pack. In Chapter 7, the integration of various battery-based technologies is described and illustrated, and comprehensive modeling, analyses, and assessments of two case studies for integrated battery-based systems for different applications are presented. The overall performances of the systems covered in the case studies are discussed, and recommendations for further improvements of the integrated battery-based systems are provided. Finally, Chapter 8 addresses challenges and future directions for the sustainable development of battery technologies in addition to providing general future considerations for battery systems.

We hope this book allows battery technologies to be applied more widely, in a sustainable manner and in a manner that facilitates sustainability of energy systems in general, using both conventional systems and renewable and advanced technologies. By applying thermodynamics, electrochemical, and thermal analyses to sustainable battery technologies, it is believed that they can be made more clean, sustainable, and efficient means for electricity storage.

Marc A. Rosen
Aida Farsi

Acknowledgments

Many people and entities assisted, directly or indirectly, in bringing this book to fruition, and we are grateful to all of them.

We particularly thank our past and present colleagues and students for their thoughtful discussions and debates, which helped influence aspects of this book. That assistance helped in enhancing the book's coverage and content and in making it more useful as a textbook and resource on battery technology and its role in sustainable energy.

We are also thankful to companies and organizations that contributed documents, data, and illustrations used in preparing this book. The materials allowed for coverage of recent developments and provided noteworthy elements of practicality as well as real-world and industrial relevance.

Last, but not least, we thank our families and friends, who have provided great motivation and support as well as understanding and patience. In particular, Marc A. Rosen thanks his wife, children, parents, siblings, and other family members, who have been a great and continuous source of inspiration and support. Furthermore, Aida Farsi warmly thanks her fiancé, Alireza, parents, and siblings, for their love, support, guidance, and motivation.

Marc A. Rosen
Aida Farsi

CHAPTER 1

Introduction to battery technology

Chapter objectives

- Describe general working principles of rechargeable battery systems.
- Identify unique vocabulary associated with battery systems and their components.
- Define and explain of basic concepts of battery technology.
- Provide characteristics and applications of various battery types.
- Identify applications of batteries and battery systems.

Nomenclature

Symbols

C_d	battery capacity in discharge (Ah)
E	electrical potential (V)
E°	standard electrical potential (V)
F	Faraday constant (96,485.3329 sA/mol)
I	battery cell current (A)
M	molecular mass (kg/kmol)
$m_{\text{electrode}}^{\text{initial}}$	quantity of electrode active materials at the initial state (kg)
$m_{\text{electrode}}^{\text{cutoff}}$	quantity of electrode active materials at the cutoff state (kg)
n_e	number of electrons
V	voltage (V)

Subscripts

a	anode
c	cathode
d	discharge

Acronyms

HVAC	heating, ventilation and air conditioning
LFP	lithium iron phosphate
LMO	lithium ion manganese oxide

Battery Technology
https://doi.org/10.1016/B978-0-443-18862-6.00002-1

NCA	nickel-cobalt-aluminum oxide
NMC	nickel-manganese-cobalt-oxide
SHE	standard hydrogen electrode

1.1 Introduction

Climate change as well as energy production and storage are topics of interest in much of the world. With increases in global population and industrialization, electricity production requirements are likely to increase, corresponding to increases in electricity demand. It is broadly hoped that this increase in electricity production can be achieved without increasing emissions of carbon dioxide and other greenhouse gases, and that electricity generation can increase with as little reliance on the fossil fuels as is reasonably possible. As a consequence, developments and advances are needed in renewable energy technologies and in energy storage technologies, especially those for storing electrical energy. Such developments and advances are important for reducing and even minimizing the impact of carbon-based fuels.

There exist numerous facilities that employ renewable energy resources like solar, wind, geothermal and tidal energies for electricity generation. Due to the intermittent nature of most types of renewable energy and the dispersed nature of these resources, most electrical power production facilities are scattered. As a consequence, the development of advanced energy storage technologies is important as a means of enabling the storage and transformation of energy for users.

Most energy storage is achieved through hydroelectric pumped storage. By 2030, the battery storage capacity is expected to reach about 600 GW (IEA, 2021). Although hydroelectric facilities are widely used for energy storage, they store only 3 Wh for every ton of water that is lifted by 1 m (Larcher and Tarascon, 2015). Therefore, other energy storage technologies need to be developed and assessed in terms of energy density, performance, durability, safety, costs (both capital and operating and maintenance), and sustainability. The term sustainability is used here in the context of battery technology to mean the technology helps meet the needs of the current generation without hindering the ability of future generations to meet their own needs.

Sources of greenhouse gas emissions in the United States for 2019, according to the Environmental Protection Agency (EPA, 2019), are shown in Fig. 1.1. It is seen that 29% of 2019 greenhouse gas emissions in the

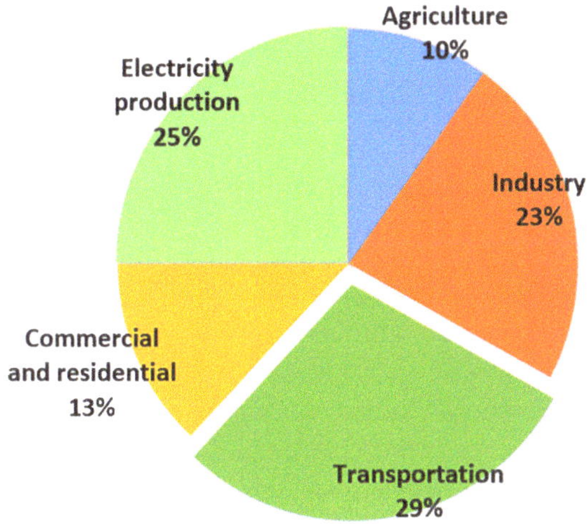

Fig. 1.1 Greenhouse gas emissions by economic sectors in 2019 in the United States. Percentage may not add up 100% due to rounding. *(Data from EPA, 2019. Sources of Greenhouse Gas Emissions. United States Environmental Protection Agency. Available from: https://www.epa.gov/ghgemissions/sources-greenhouse-gas-emissions (Accessed 25 May 2022).)*

United States were attributable to the transportation sector (see Fig. 1.1). Fossil fuel use for driving trains, airplane, ships, cars, and trucks are the main contributors to greenhouse gas emissions associated with transportation. Road transportation, in particular, is a key contributor to greenhouse gas emissions by the transportation sector. Governmental regulations for environment protection, especially those addressing climate change and global warming, have fostered increasing global interest in electric vehicles and hybrid electric vehicles. This increases the importance the improved design and development of batteries as an electrical energy storage in general, and especially for electric and hybrid electric vehicles. Efficient batteries can potentially reduce greatly emissions of carbon dioxide and other greenhouse gases.

In recent years, significant progress has been achieved in battery technologies, and battery usage has increased significantly and expanded in breadth of application. Nonetheless, there remain challenges regarding battery performance and cost that need to be addressed for broader market infiltration. Modifications of existing battery designs and the usage of advanced materials in fabrication are key elements in the advancement of battery technology.

In this chapter, an introduction to and overview of battery technology is presented, including descriptions of general working principles as well as the characteristics of and materials used for various types of commercial batteries. Applications of different battery types are presented, and their characteristics like specific energy density, lifetime and cost are compared.

1.2 Battery working principles

Batteries are devices that store energy in the form of chemical energy and convert it when required to electrical energy. Through the electrochemical reactions that occur in a battery, electrons are released and flow from one conductor (electrode) to another through external electric circuit providing an electric current that is used to do tasks. At the same time, charged ions are transported through an electrically conducting solution (electrolyte), which is in contact with the electrodes, to bring the reactants to the electrode/electrolyte interface.

The electrodes and electrolyte can be made from various materials. Variations in the composition of the electrolyte and electrodes result in different electrochemical reactions and charged ions. These influence the extent to which energy can be stored, as well as the operating voltage and performance of the battery.

A battery can be comprised of one or more electrochemical cells. Each electrochemical cell includes two electrodes which are separated by an electrolyte. During discharge, an electrode from which electrons flow is known as *anode* or *negative electrode* and an electrode which receives the electrons is known as *cathode* or *positive electrode*. Typically, the anode and cathode are made of different types of chemical compounds or metals. *The electrolyte* is a medium that allows the charged ions be transported toward the electrodes. In fact, the flow of positive charged ions through the electrolyte balances the movement of negative electrons which makes the battery cell electrically neutral during its operation. Also, a porous and electronically insulating *separator* is often used between the anode and cathode to improve the mechanical strength of the electrolyte and reduce the risk of an internal short circuit. Typically, separators have a high ionic conductivity and provide electronic insulation. The *current collectors* also can be used to provide efficient transfer of electrons and to remove heat from the electrodes. Thin foils of aluminum or copper are typically used as current collectors.

A reaction in which electrons are exchanged is referred to a reduction-oxidation reaction or redox reaction. The overall reaction is divided into two half reactions; in an electrochemical cell, one half reaction take places at the cathode and the other at the anode. A general redox reaction can be written as follows:

$$O + n_e e^- \rightleftharpoons R \tag{1.1}$$

where O denotes the oxidized species and R the reduced species. Reduction refers to the gain of electrons (which happens at the cathode). The cathode is said to be reduced during the reaction. Oxidation refers to the loss of electrons (which happens at the anode), and the anode is said to be oxidized during the reaction. So, the reaction at the anode is oxidation and at the cathode is reduction. These reactions have their own standard potentials, which indicate the capability of the reactions for accepting or producing electrons.

Fig. 1.2 shows a basic operational mechanism for an electrochemical battery cell. Each cell contains two electrodes and at least one electrolyte. During discharge of the battery, at the anode/electrolyte interface, an electrochemical reaction of $R \rightarrow O + n_e e^-$ occurs between the anode and the electrolyte that produces electrons and ions. The electrons are transferred to the cathode through an external electrical circuit which connect the electrodes, and the ions are transferred to the cathode through an electrolyte. The electrons and ions that were produced in the anodic half-reaction react at the cathode/electrolyte interface (i.e., $O + n_e e^- \rightarrow R$).

1.2.1 Electrodes

Any two electrodes that have different standard reaction potentials can create an electrochemical cell. The electrode with higher ability to absorb electrons (as happens at the cathode) has a higher standard potential (i.e., more positive potential) while the electrode with a higher tendency to lose electrons

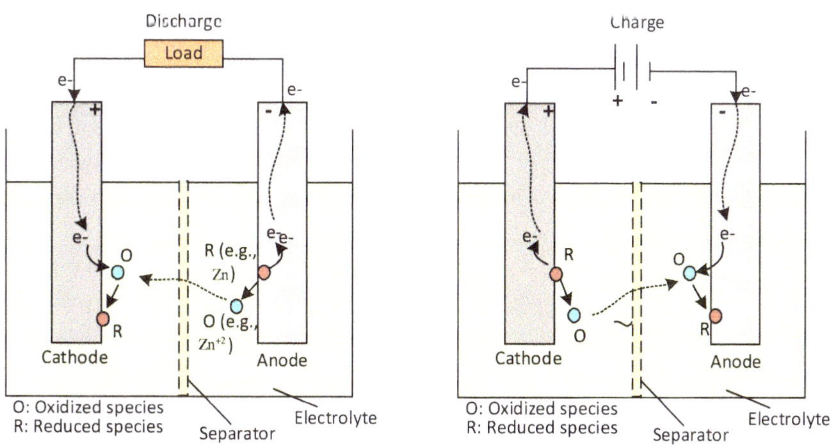

Fig. 1.2 Generalized redox battery operation during discharge and charge.

(as happens at the anode) has a lower standard potential (i.e., more negative potential). Therefore, an appropriate selection of materials for the anode and the cathode is based on more negative and more positive standard reaction potentials, respectively.

Typically, group 1 and 2 metals in the periodic table are easily oxidized to A^+ or A^{2+} (A is a general representative for group 1 and 2 metals). Also, group 16 and 17 metals in the periodic table are easily reduced to B^- or B^{2-} (B is a general representative for group 16 and 17 metals). In addition, the lighter the material, the higher is the energy density and subsequently the better it is in the fabrication of electrodes. Lithium (Li), magnesium (Mg), and sodium (Na) are the lightest and most easily oxidized metals, and oxygen (O) and sulfur (S) are the lightest and most easily reduced elements. Although halogens are easier oxidized compared to Li, Mg, and Na, the use of these elements in the construction of an anode is not practical.

The desired characteristics for the cathode are high redox potential (efficient oxidizing agent), high specific capacity (total amount of electricity generated in the electrochemical reactions per gram of cathode material), reversibility and stability while in contact with the electrolyte. For the anode, the desired characteristics are low redox potential (efficient reducing agent), high specific capacity, reversibility and good conductivity. Table 1.1 lists several important parameters and their values for several anode and cathode materials used in different types of battery cells, including molecular weight, valence, standard reduction potential, gravimetric capacity, and volumetric capacity.

The difference between the electrical potentials of the cathode and the anode gives the cell potential, i.e.,

$$E^{\circ}_{\text{cell}} = E^{\circ}_{\text{cathode}}(\text{reduction potential}) - E^{\circ}_{\text{anode}}(\text{oxidation potential}) \quad (1.2)$$

The greater the difference, the greater is the cell potential, and the higher is the voltage. A practical way to measure the electrical potential of an electrode (cathode or anode) is to assign zero for the electrical potential of the reaction that occurs at the electrode and then use it as a reference electrode.

For example, at room temperature, consider a platinum electrode inserted in a 1 molar concentration (M) hydrochloric acid solution. Molar concentration is a measure of the concentration of a solute in a solution, in terms of amount of substance per unit volume of solution. A 1 M hydrochloric acid solution indicates that there is 36.46 g HCl dissolved in 1 L of water. Hydrogen gas at 1 atm is bubbled from the platinum electrode that is inserted

Table 1.1 Characteristics of selected electrode materials.

Material	Molecular or atomic weight (g)	Standard redox potential at 25°C (V)	Valence	Gravimetric capacity (Ah/g)	Volumetric capacity (Ah/cm^3)
Anode					
H_2	2.01	0	2	26.59	–
Li	6.94	−3.01	1	3.86	2.06
Na	23	−2.71	1	1.16	1.14
Mg	24.3	−2.38	2	2.20	3.8
Al	26.9	−1.66	3	2.98	8.1
Ca	40.1	−2.84	2	1.34	2.06
Fe	55.8	−0.44	2	0.96	7.5
Zn	65.4	−0.76	2	0.82	5.8
Cd	112.4	−0.4	2	0.48	4.1
Pb	207.2	−0.13	2	0.26	2.9
LiC_6	72.06	−2.8	1	0.37	0.84
MH	116.2	−0.83	2	0.45	–
Cathode					
O_2	32.0	1.23	4	3.35	–
Cl_2	71.0	1.36	2	0.756	–
SO_2	64.0	–	1	0.419	–
MnO_2	86.9	1.28	1	0.308	1.54
NiOOH	91.7	0.49	1	0.292	2.16
CuCl	99.0	0.14	1	0.270	0.95
FeS_2	119.9	–	4	0.89	4.34
AgO	123.8	0.57	2	0.432	3.20
Br_2	159.8	1.07	2	0.335	–
HgO	216.6	0.10	2	0.247	2.74
Ag_2O	231.7	0.35	2	0.231	1.64
PbO_2	239.2	1.69	2	0.224	2.11
Li_xCoO_2	98	−2.7	0.5	0.137	–
I_2	253.8	0.54	2	0.211	1.04

Data from Lide, D.R. (Ed.), 2004. CRC Handbook of Chemistry and Physics, 85th ed. CRC Press, Boca Raton, FL, USA.

in the 1 M hydrochloric acid solution. The reduction half-reaction is considered as the reference, that is,

$$2H^{+}(aq,\ 1\ M) + 2e^{-} \rightleftharpoons H_2(g,\ 1\ atm) \quad E^{\circ} = 0 \tag{1.3}$$

This reference electrode is called the standard hydrogen electrode (SHE). Hydrochloric acid solution (HCl (aq)) is used because platinum is inert to the

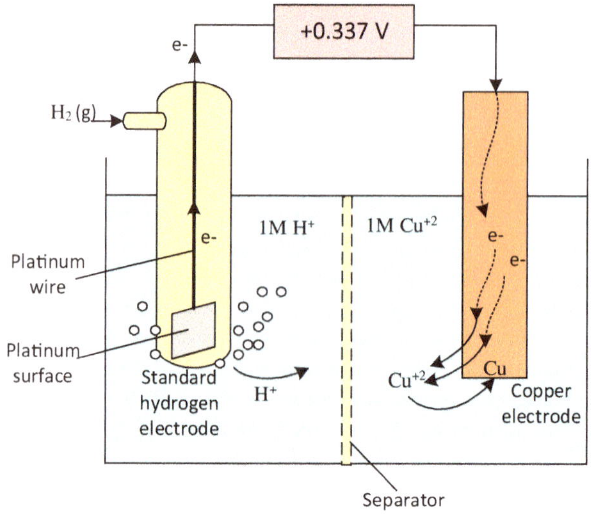

Fig. 1.3 An electrochemical cell including copper and standard hydrogen electrodes. Hydrogen gas at 1 atm is bubbled in a 1 M HCl (aq) solution.

action of the 1 M HCl, The reason for the selection of platinum is that it is inert to the action of 1 M hydrochloric acid (HCl). Electrons on the surface of the platinum electrode react with the proton (i.e., H^+) to produce hydrogen gas (H_2). Consider an electrochemical cell shown, as shown in Fig. 1.3, including copper and SHE half cells in which reduction occurs at the copper electrode and oxidation at the SHE. That is:

$$\text{Cathode (reduction)} : Cu^{2+}(aq) + 2e^- \rightarrow Cu(s) \quad E^{\circ}_{cathode} \tag{1.4}$$

$$\text{Anode (oxidation)} : H_2(g) \rightarrow 2H^+(aq) + 2e^- \quad E^{\circ}_{anode} \tag{1.5}$$

$$\text{Overall} : Cu^{2+}(aq) + H_2(g) \rightarrow 2H^+(aq) + Cu(s) \quad E^{\circ}_{cell} \tag{1.6}$$

As mentioned previously, the standard cell potential is equal to the difference between the standard reaction potentials at the anode and cathode, as follows:

$$E^{\circ}_{cell} = E^{\circ}_{cathode} - E^{\circ}_{anode} \tag{1.7}$$

Since the electrical potential for the reaction that occurs at the SHE is zero (i.e., E°_{anode}), the displayed value on a voltmeter (which shows the cell electrical potential) is equivalent to the standard electrical potential for the reduction reaction at the copper electrode (i.e., $E^{\circ}_{cell} = E^{\circ}_{cathode} = +0.337$ V).

1.2.2 Electrolyte

The electrolyte can be a liquid, solid, polymer, or composite (hybrid) depending on the type of battery. The electrolyte should have a high ionic conductivity, no electric conductivity, nonreactivity with the electrode materials and a wide operating temperature range. The traditional *liquid electrolyte* generally has a low viscosity, high energy density, high charge/discharge rate capability, a relatively low operational temperature (between $-40°C$ and $60°C$) and low flammability. The *polymeric electrolyte* can be either a gel or a solid. The *solid polymeric electrolyte* has the advantageous of high flexibility, high energy density, multifunctional applications, good safety and mechanical properties and thermal/chemical stability. But solid polymeric electrolytes typically have low ionic conductivities at room temperature (10^{-5}–10^{-1} mS cm^{-1}). *Gel polymeric electrolytes*, on the other hand, have relatively high ionic conductivities (1 mS cm^{-1}), high flexibility, multifunctional applications, and chemically stability, but they have poor mechanical strength and poor interfacial properties. The main advantages of a *solid polymeric electrolyte* are no electrolyte leakage, high safety (nonflammability), nonvolatility, thermal and mechanical stability, ease of fabrication, and high achievable power density and cyclability.

The voltage of a battery can be increased through either selecting materials for the anode and the cathode to yield a greater cell potential, or by attaching or stacking multiple cells together. The arrangement of battery cells in a module can be in parallel or in series. In the parallel arrangement, the battery current increases due to an increase in the total number of electrons flowing through the battery. For a given circuit, from each battery a set number of electrons can be drawn per second, so if two or more batteries are connected in parallel, the number of electrons they can transport per second is multiplied and, subsequently, the total current in the circuit is increased.

In the series arrangement of battery cells, the battery voltage increases as the force to move the electrons through the battery increases. For instance, when two identical batteries are connected in series, the voltages are added because since the same charge is moved twice each time through the same voltage. Thus, the total work done is doubled (i.e., 2 V), while the current remains constant.

Furthermore, battery cells can be connected in a series-parallel arrangement to increase the voltage and current of battery system at the same time. The total current (I_T) and the total voltage (V_T) for a series-parallel arrangement can be determined as:

$$I_T = \sum_{n=\text{number of rows}} I_n \tag{1.8}$$

$$V_T = \sum_{m=\text{number of columns}} V_m \tag{1.9}$$

where the rows are connected in series and the columns in parallel.

As an example, consider the latter case of a series-parallel arrangement. If eight battery cells with characteristics of voltage 6 V and current 100 Ah are connected in two rows, each of which includes four cells, the resulting voltage and current for this module would be:

$$I_T = \sum_{n=2} I_n = I_1 + I_2 = 2I_1 = 2 \times 100 \text{ Ah} = 200 \text{ Ah} \tag{1.10}$$

$$V_T = \sum_{m=4} V_m = V_1 + V_2 + V_3 + V_4 = 4(V_1) = 4 \times 6 \text{ V} = 24 \text{ V} \tag{1.11}$$

Fig. 1.4 shows the connections of battery cells in a module for this example.

1.2.3 Battery capacity

The capacity of a battery is equivalent to the amount of electric charge that is accumulated during the charge mode, then stored during the open circuit stay and finally used during the discharge mode in a reversible manner. The capacity of a battery can be determined by integrating of discharge current between the limits of a completely charged battery and a discharge state at the voltage threshold (also known as cutoff voltage, V_{cutoff}). The

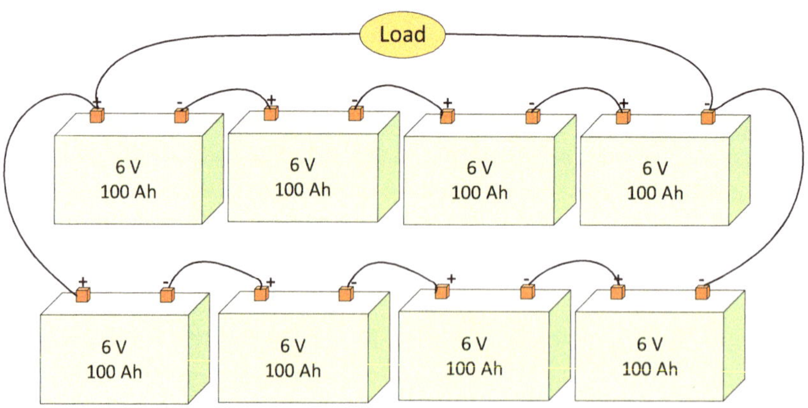

Fig. 1.4 A parallel-series arrangement of battery cells for delivering high voltage and current. Shown are two parallel rows, each with four cells in series.

corresponding time at this voltage is cutoff time (t_{cutoff}). When the battery is discharged, its discharge capacity (C_d) is given by:

$$C_d = \int_{\text{initial}=0}^{t_{\text{cutoff}}} I dt = \frac{-n_e F}{M_{\text{electrode}}} \left(m_{\text{electrode}}^{\text{initial}} - m_{\text{electrode}}^{\text{cutoff}} \right) \tag{1.12}$$

where $M_{\text{electrode}}$ denotes the molecular weight of the electrode material, I the cell current, n_e the number of electrons, and F the Faraday constant, while $m_{\text{electrode}}^{\text{initial}}$ and $m_{\text{electrode}}^{\text{cutoff}}$ are the quantities of electrode active materials at the initial state (completely charged battery) and cutoff state, respectively. In an ideal condition where there is no capacity fade due to irreversibilities within the battery, Eq. (1.12) can be expressed for either the positive or the negative electrode material, as the battery releases the same electric charge during discharge as it receives during charge. For instance for a lead acid battery, C_d can be given by:

$$C_d = \int_{\text{initial}=0}^{t_{\text{cutoff}}} I dt = \frac{-2F}{M_{\text{PbO}_2}} \left(m_{\text{PbO}_2}^{\text{initial}} - m_{\text{PbO}_2}^{\text{cutoff}} \right) = \frac{-2F}{M_{\text{Pb}}} \left(m_{\text{Pb}}^{\text{initial}} - m_{\text{Pb}}^{\text{cutoff}} \right) \tag{1.13}$$

During discharge, the active materials in the electrodes are converted to discharge products through electrochemical reactions until the voltage of the battery approaches the voltage threshold. The extent of this conversion is known as the *utilization of electrode material factor*.

In some batteries, a reversible reaction can occur in the cell, which allows for recharging of the battery. In fact, with appropriate selection of materials for electrodes, both forward and backward movement of electrons through external circuits is possible. The electrochemical reactions that occur at the anode and cathode during discharge are reversed during charging of a battery. These kinds of batteries are known as secondary batteries, and they can be charged and discharged many times before reaching the ends of their useful lifetimes.

The primary or disposable batteries, on the other hand, are referred to batteries that the electrons can move only from anode to cathode during discharge. In other words, the reactions in anode and cathode cannot occur in reverse direction. This type of battery stores and delivers energy only one time. In primary batteries, the lifetime of the battery ends when the electrodes are depleted as they release all ions to the electrolyte or the products completely cover the electrode surface and do not allow any more reactants to reach the interface, thus preventing the reaction from occurring any more.

This book is mainly focused on secondary batteries. It should be noted that for convenient throughout the book, the terminology of "battery" is used for "rechargeable battery." Since secondary batteries are rechargeable, the forward and backward movement of ions through an electrolyte and the reactions at the interfaces are not perfect compared to the first discharge process. This means that the performance of such a battery degrades over the charge and discharge cycles and it eventually approaches its end of lifetime.

1.3 Types of batteries

There are numerous types of batteries. These differ in terms of cell components such as electrolyte, anode, and cathode. Different types of batteries have different operating temperature ranges, fabrication costs, specific energy (energy content per weight of battery) values, specific power (loading capability) values, efficiencies and life cycles and lifetimes. An important goal in battery technology development is to improve these values.

In the following subsections, various types of secondary batteries that are commercialized are presented and described.

1.3.1 Lead acid batteries

Lead acid batteries are comprised of an aqueous sulfuric acid electrolyte, a lead anode, and a lead dioxide cathode. During discharging, the anode and cathode consume sulfuric acid present in the electrolyte, and convert it to lead sulfate. During charging, the lead sulfate is converted back to sulfuric acid through depositing a layer of metallic lead on the anode and lead oxide on the cathode.

Fig. 1.5 demonstrates the working principle of a lead acid battery. During discharge, when the load is externally connected between the electrodes, the molecules of the sulfuric acid (H_2SO_4) electrolyte are split into positive hydrogen ions (H^+) and negative sulfate ions (SO_4^{2-}). The positive hydrogen ions reach the lead peroxide electrode (cathode) and receive electrons from this electrode. The hydrogen ions then become hydrogen atoms, which again react with lead peroxide and form lead oxide and water. The produced lead oxide reacts with sulfuric acid and forms lead sulfate and water. Some of the negative sulfate ions move toward the lead electrode (anode) to which they transfer their extra electrons and become sulfate radicals. The sulfate radicals react with pure lead and form lead sulfate. As the positive hydrogen ions receive electrons from the cathode and the negative sulfate ions give their electrons to the anode, there remains an inequality of electrons

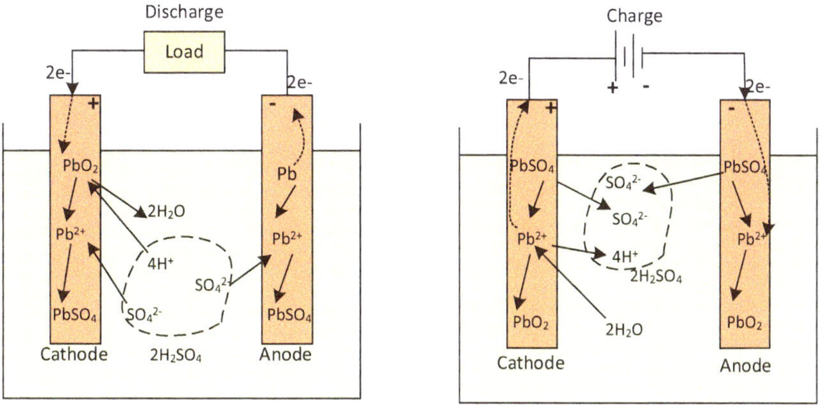

Fig. 1.5 Working principle of lead acid battery cell during charge and discharge.

between the anode and cathode. Therefore, there is a flow of electrons (current) through the external load between these two electrodes, which balances this inequality of electrons.

The redox reactions at the anode and cathode of lead acid batteries during discharge can be written as follows:

$$\text{Oxidation in anode}: Pb\,(s) + SO_4^{2-}(aq) \rightarrow PbSO_4(s) + 2e^-; \quad E_a^\circ = -0.356\ V \tag{1.14}$$

$$\text{Reduction in cathode}: PbO_2(s) + 4H_3O^+(aq) + SO_4^{2-}(aq) + 2e^- \rightarrow PbSO_4(s) + 6H_2O(I); E_c^\circ = 1.685\,V \tag{1.15}$$

$$\text{Overall reaction in cell}: Pb\,(s) + PbO_2(s) + 2H_2SO_4(aq) \rightarrow 2PbSO_4(s) + 2H_2O(I); E_{cell}^\circ = E_c^\circ - E_a^\circ = 2.05\,V \tag{1.16}$$

At high cell voltages, as a result of the low potential of the anode and the high potential of the cathode, a number of secondary reactions can occur. These can include hydrogen evolution at the anode ($4H^+ + 4e^- \rightarrow 2H_2$) and oxygen evolution at the cathode ($2H_2O \rightarrow 4H^+ + 4e^- + O_2$), which yield an overall reaction of water decomposition. In fact, water molecules in the electrolyte are dissociated into oxygen and hydrogen gases during the charging phase. This is referred to as water electrolysis which is an undesired reaction in lead acid batteries. If the battery is designed so as to vent these gases to the atmosphere, makeup water is added to the battery system to compensate for the venting. Alternatively, in a sealed design of battery, the hydrogen and oxygen gases are trapped and converted to water. Although the sealed design of lead acid batteries eliminates the periodic addition of water to the battery cell, during conditions with high rates of charge or overcharge, the generation rate of hydrogen and oxygen gases can exceed the production rate of water (i.e., the rate of combination of the gases). This may increase the risk of explosion in a lead acid battery cell. Also, self-discharge of the battery may occur. This involves electrons being transferred from the anode to the cathode internally through the electrolyte rather than through an external circuit.

Lead acid batteries are a relatively low cost type of battery. However, they generally have low specific energies (amount of energy that can be delivered from a battery in comparison to its weight) and the lead they contain is toxic. In practice, the specific energy of a lead acid battery is between 30 and 40 Wh/kg of battery, while the average theoretical value is about 170 Wh/kg. This range of specific energy for lead acid batteries is significantly lower than that for lithium ion batteries, which typically have average specific energies of about 250 Wh/kg.

The useful lifetimes of lead acid batteries are usually only a few hundred cycles. Shedding of active material from the surface of the electrode and corrosion are two of the primary reasons for relatively short lifetimes of lead acid batteries. The accumulation of sulfate on the surface of the anode due to high temperature and high rate of operations is an additional significant reason for degradation in lead acid batteries. There are several ways to extend the lifetimes of lead acid batteries. These include:

(i) improving the grid design and active materials in the electrodes,
(ii) using a carbon anode instead of a lead anode to store electrostatic charge, and
(iii) using a lead acid flow battery in which lead is dissolved in a methane sulfonic acid electrolyte.

These methods (specifically the last one) also can potentially enhance the specific energy of lead acid batteries somewhat, although they still have relatively low specific energies compared to other types of batteries.

1.3.2 Nickel-based batteries

Nickel-based batteries include various types, including nickel-cadmium (Ni-Cd), nickel-iron (Ni-Fe), nickel-zinc (Ni-Zn), and nickel-hydrogen (Ni-H_2). In all of these nickel-based batteries, a nickel hydroxide electrode is used as the cathode. Since a strong alkaline solution (like potassium hydroxide) is usually used as the electrolyte in the nickel-based batteries, they are also known as alkaline secondary batteries. The difference between the types of nickel-based batteries is linked to the material used for the anode.

The charge and discharge reactions for a nickel electrode (the cathode for all types of nickel-based batteries) can be expressed as follows:

$$\beta \text{NiOOH} + H_2O + e^- \underset{\text{charge}}{\overset{\text{discharge}}{\rightleftarrows}} \beta \text{Ni(OH)}_2 + OH^-;\ E_c^\circ = 0.49\ \text{V} \tag{1.17}$$

where β is the stoichiometry coefficient, which takes on different values for different types of nickel-based batteries.

In the following subsections, five types of nickel-based batteries are described. These include nickel-cadmium, nickel-iron, nickel-hydrogen, nickel-metal hydride, and nickel zinc batteries.

1.3.2.1 Nickel-cadmium batteries

This type of nickel-based battery consists of a nickel (III) oxide-hydroxide material as the cathode, a cadmium plate as the anode, an alkaline electrolyte (usually potassium hydroxide) and a separator. The electrolyte is usually maintained at a constant concentration (e.g., 20%–30% KOH concentration) during the charge-discharge cycle to minimize the cell resistance.

The redox reactions during discharge at the anode and cathode of nickel-cadmium batteries can be written as follows:

$$\text{Oxidation at anode}: \text{Cd} + 2\text{OH}^- \rightarrow \text{Cd(OH)}_2 + 2e^-;\ E_a^\circ = -0.81\ \text{V} \tag{1.18}$$

$$\text{Reduction at cathode}: 2\text{NiOOH} + 2H_2O + 2e^- \rightarrow 2\text{Ni(OH)}_2 + 2\text{OH}^-;\ E_c^\circ = 0.49\ \text{V} \tag{1.19}$$

$$\text{Overall reaction}: 2\text{NiOOH} + \text{Cd} + 2H_2O \rightarrow 2\text{Ni(OH)}_2 + \text{Cd(OH)}_2;\ E_{cell}^\circ = 1.3\ \text{V} \tag{1.20}$$

During the charging of a Ni-Cd battery, the above anodic, cathodic, and overall reactions take place in the reverse directions.

Fig. 1.6 shows the working principles of a Ni-Cd battery cell during charge and discharge. During discharge, the nickel oxide hydroxide combines with water to form nickel hydroxide and hydroxide ions at the

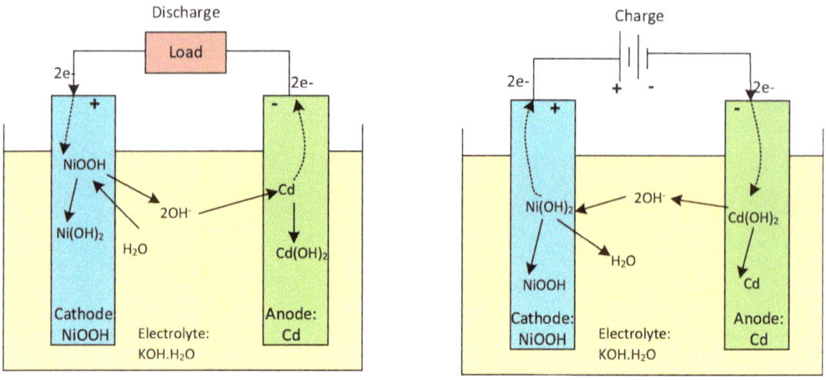

Fig. 1.6 Working principle of nickel-cadmium battery cell during discharge and charge.

cathode. Metallic cadmium is oxidized to produce cadmium hydroxide and two electrons at the anode. During this process, the battery cell liberates energy. The process is reversed during charging, but the oxygen is released at the positive electrode and hydrogen is formed at the negative electrode. So, a quantity of water needs to be added to compensate for the hydrogen and oxygen lost during the decomposition of water.

A Ni-Cd battery has a nominal cell potential of 1.3 V. Ni-Cd batteries are used for wide range of electric devices due to their relatively high energy densities (50–75 Wh/kg) and lifetimes (2000–2500 charge/discharge cycles). However, their usage is limited due to the toxicity of cadmium and the high cost.

The performance of a Ni-Cd battery is highly affected by operating temperature, rate of charge/discharge, cell type and construction and manufacturing process. For example, Fig. 1.7 shows the effect of temperature on the Ni-Cd cell voltage. It is seen that, in a fully charged cell (i.e., 100% for the ratio of charge input to the percentage of cell capacity), the cell voltage in a Ni-Cd cell decreases from 1.53 to 1.37 V as the operating temperature increases from 15°C to 40°C.

1.3.2.2 Nickel-iron batteries

In this type of Ni-based battery, iron is used for the anode and nickel oxyhydroxide for the cathode. These are immersed in an alkaline solution (e.g., 30% KOH concentration) as the electrolyte. The redox reactions during discharge for a nickel-iron battery can be written as follows:

$$\text{Oxidation in anode}: Fe + 2OH^- \rightarrow Fe(OH)_2 + 2e^-; E_a^\circ = 0.88\ V \tag{1.21}$$

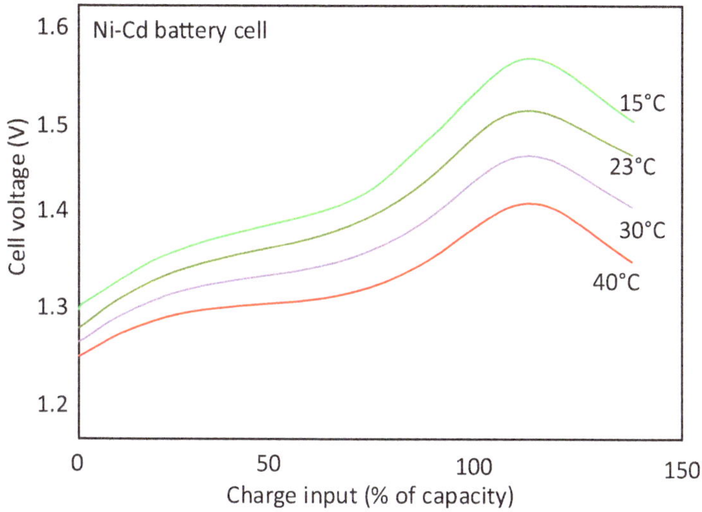

Fig. 1.7 Effect of operating temperature and level of charge on the voltage of nickel-cadmium cell battery. *(Data from Shukla, A.K., Venugopalan, S., Hariprakash, B., 2001. Nickel-based rechargeable batteries. J. Power Sources 100(1–2), 125–148.)*

$$\text{Reduction in cathode}: 2NiOOH + 2H_2O + 2e^- \rightarrow 2Ni(OH)_2 + 2OH^-; E_c^\circ = 0.49\ \text{V} \quad (1.22)$$

$$\text{Overall reaction}: 2NiOOH + Fe + 2H_2O \rightarrow 2Ni(OH)_2 + Fe(OH)_2; E_{cell}^\circ = 1.37\ \text{V} \quad (1.23)$$

The energy density of Ni-Fe battery cells is typically between 19 and 25 Wh/kg. A vent can be used in a Ni-Fe cell to release the hydrogen gas produced in the cell. Hydrogen is produced because, when electricity passes through the anode and cathode of a nickel-iron battery, nickel oxide hydroxide and reduced iron are produced. These serve as catalysts for the decomposition of water.

The lifecycles of typical Ni-Fe battery cells are about 3000 cycles, with the lifetime of 20 years at operating temperatures lower than 30°C. However, the useful lifecycle and lifetime of Ni-Fe batteries are affected by operating conditions like cell temperature and rate of charge/discharge. For instance, at relatively high operating temperatures (about 40°C), the battery lifecycle and lifetime of Ni-Fe cells are only 1500 cycles and 8 years, respectively.

Several corrosion reactions may occur in Ni-Fe cells. These reactions can lead to self-discharge of about 1%–2% of the cell's nominal capacity at room temperature. The phenomenon of self-discharge is usually intensified in hot weather conditions (about 40°C). Hydrogen evolution may also occur during the charging process of Ni-Fe cells, which reduces the charge acceptance by the cell. Using an electro-catalyst for the iron electrode to reduce self-discharge and applying a hydrogen-oxygen recombination catalyst to recombine evolved hydrogen and oxygen may improve the performance of Ni-Fe cells. Nonetheless, such cells are still not able to compete with other types of batteries and need further enhancements to raise commercial interest.

1.3.2.3 Nickel-hydrogen batteries

In nickel-hydrogen (Ni-H_2) cells, a hydrogen gas electrode is used as an anode and is stored in a pressurized cell. A 26% potassium hydroxide (KOH) solution is typically used as an electrolyte in the Ni-H_2 cells.

The development of Ni-H_2 cells was inspired from both battery and fuel cell technologies. Specifically, the nickel hydroxide electrode is drawn from Ni-based battery cells and the hydrogen electrode from fuel cell technology. The relatively light-weight of the hydrogen electrode enhances the cell gravimetric energy density (about 75 Wh/kg), but lowers the volumetric energy density (between 60 and 100 Wh/L) compared to other nickel-based cells. This type of nickel-based battery is often used for onboard energy conversion and storage in aerospace applications. Nickel-hydrogen battery cells can tolerate overcharging. The distinctive feature of Ni-H_2 battery cells is their long lifecycle, which exceed 20,000 charge-discharge cycles, and high energy efficiencies (around 85%). However, their high costs prevent their use in other applications than aerospace.

The redox reactions at the anode and cathode of Ni-H_2 cells can be written as follows:

$$\text{Oxidation in anode}: 0.5\text{H}_2 + \text{OH}^- \rightarrow \text{H}_2\text{O} + \text{e}^-; E_a^\circ = -0.83\ \text{V} \tag{1.24}$$

$$\text{Reduction in cathode}: 2\text{NiOOH} + 2\text{H}_2\text{O} + 2\text{e}^- \rightarrow 2\text{Ni(OH)}_2 + 2\text{OH}^-; E_c^\circ = 0.49\ \text{V} \tag{1.25}$$

$$\text{Overall reaction}: \text{NiOOH} + 0.5\text{H}_2 \rightarrow \text{Ni(OH)}_2; E_{cell}^\circ = 1.32\ \text{V} \tag{1.26}$$

During discharge, at the cathode, the hydrogen electrode inside the pressure vessel is oxidized to water, whereas at the anode, the nickel oxyhydroxide electrode is reduced to nickel hydroxide. The concentration of potassium hydroxide electrolyte does not significantly change as the water that is used at the anode (nickel electrode) is reproduced at the cathode (hydrogen electrode).

Various designs for Ni-H_2 battery cells have been developed and these are comprehensively described elsewhere (Shukla et al., 2001). As an example, Fig. 1.8 shows a basic configuration of a common pressure vessel Ni-H_2 battery cell developed by COMSAT laboratory (NASA, 1997). Alternating pairs of anode (hydrogen electrode) and cathode (nickel electrode) are separated by a zircar (a combination of zinc peroxide and radical anion) separator that is wetted by the electrolyte (i.e., potassium hydroxide solution). All cell components are stacked in a cylindrical pressure vessel, usually a hermetically sealed Inconel (a solid solution strengthened Ni-based superalloy) that allows for operation between 3.45 bar and 69 bar. The pressure vessel is typically designed to have a burst strength that is three times the maximum operating pressure (Shukla et al., 2001). The hydrogen electrode (anode) is comprised of a Teflon bonded platinum black

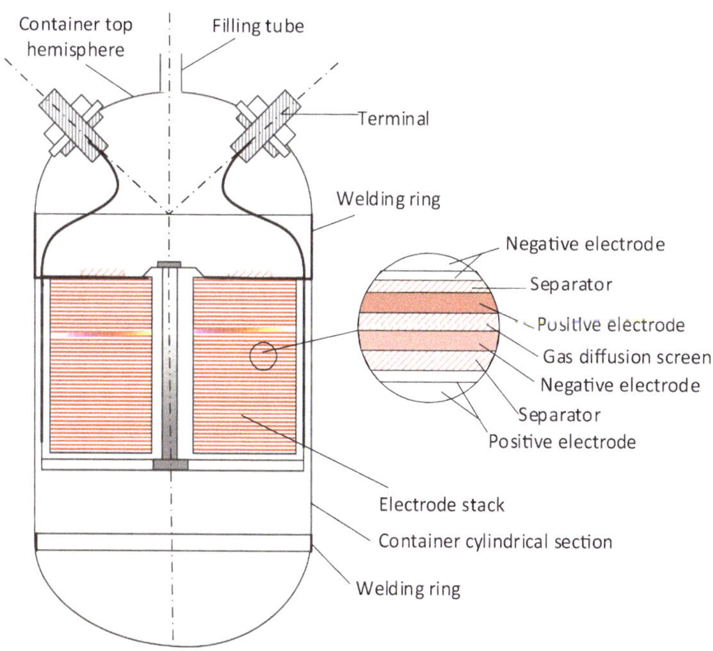

Fig. 1.8 Common pressure vessel nickel-hydrogen battery positioned vertically. *(Redrawn from NASA, 1997. Lesson Number 568. The National Aeronautics and Space Administration. Available from: https://llis.nasa.gov/lesson/568 (Accessed 25 May 2022).)*

catalysis on a nickel screen with a porous gas-diffusion Teflon backing film. The nickel electrode (cathode) is made up of dry sintered porous nickel plates that are coated with a nickel hydroxide active material.

1.3.2.4 Nickel-metal hydride batteries

In nickel-metal hydride (Ni-MH) battery cells, a complex metal alloy is used for the anode and nickel with NiOOH as an active material for cathode. The complex metal alloy in the anode serves as a solid source of reduced hydrogen that can be oxidized to generate protons. A KOH solution is used as the electrolyte. Ni-MH systems were initially designed for storage of hydrogen in nickel hydrogen cells. This type of battery cell has a moderate specific energy (70–100 Wh/kg) and relatively high volumetric energy density (170–420 Wh/L), and can tolerate overdischarge. Also, Ni-MH cells are free from dendrite formation and, unlike Ni-Cd, Ni-Fe and lead acid batteries (which contain some toxic substances), no toxic material is used in them.

Fig. 1.9 shows the operating principle of a nickel-metal hydride battery cell. During discharge, the hydroxide ions are generated from combination of water molecules and nickel-oxide hydroxide at the cathode. The generated hydroxide ions move from cathode to anode through the KOH solution. Hydroxide ions transferred to the anode receive hydrogen ions from the hydrogen storage alloy and return to the water molecules.

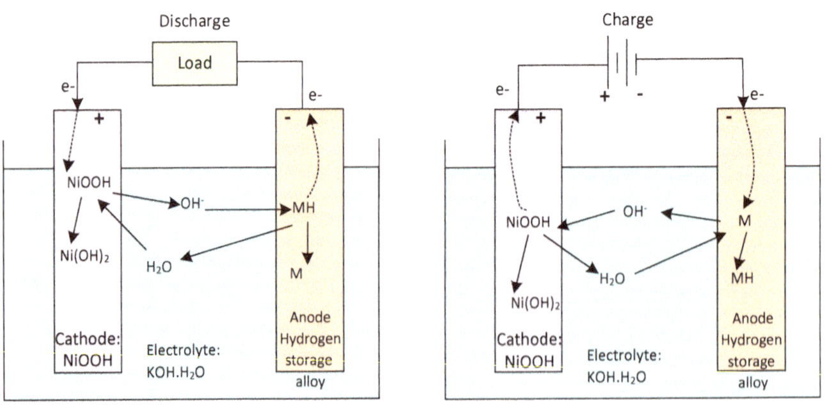

Fig. 1.9 Working principle of nickel-metal hydride battery cell during discharge and charge.

The redox reactions during discharge of Ni-MH battery cells can be written as follows:

$$\text{Oxidation at anode}: MH + OH^- \rightarrow M + H_2O + e^-; E_a^\circ = -0.83\text{ V} \tag{1.27}$$

$$\text{Reduction at cathode}: NiOOH + H_2O + e^- \rightarrow Ni(OH)_2 + OH^-; E_c^\circ = 0.49\text{ V} \tag{1.28}$$

$$\text{Overall reaction}: NiOOH + MH \rightarrow M + Ni(OH)_2; E_{cell}^\circ = 1.32\text{ V} \tag{1.29}$$

The directions of the oxidation, reduction and overall reactions written above are reversed during the charge process of Ni-MH battery cells.

The operation principle of a Ni-MH battery cell is based on its ability to absorb, release and transport hydrogen between the anode and the cathode. During discharge, hydrogen that is stored in the metal alloy on the anode is released into the hydroxide potassium electrolyte to form water. The produced water is dissociated into a hydrogen ion that is absorbed by the cathode to form nickel hydroxide ($Ni(OH)_2$). The process of transporting hydrogen from the anode to the cathode is endothermic (i.e., it takes in heat). The absorption of heat continues until the cell is overdischarged. At this state, a secondary reaction occurs that leads to an increase in the temperature of Ni-MH battery cell. So, due to endothermic discharge reactions, a high discharge rate with relatively low heat dissipation is observed for Ni-MH battery cells.

The metal alloy in which hydrogen is stored at the anode is often a Misch metal. A Misch metal is an alloy of rare-earth elements with iron and traces of sulfur, carbon, calcium, and aluminum. Typically, the composition is about 50% cerium, 25% lanthanum, 15% neodymium, and 10% other rare-earth metals and iron. This metal alloy has a high volumetric energy density, which results in a high battery capacity.

The advantages of Ni-MH battery cells, like good cycle life, high rate capability and high specific energy, is leading to a tendency for replacing Ni-Cd with Ni-MH batteries. However, further improvements are required to improve the high self-discharge rates (up to 30% per month) of Ni-MH cells and their performance at relatively low temperatures (i.e., below 0°C).

Fig. 1.10 shows the general components of a cylindrical nickel-metal hydride battery cell, including the cathode and the anode separated by an electrolyte, as well as the separator and casing. The electrolyte, electrode and separator are rolled in a steel case that serves as a sealing plate. In addition, a safety valve is used to avoid bursting occurred due to a short circuit or

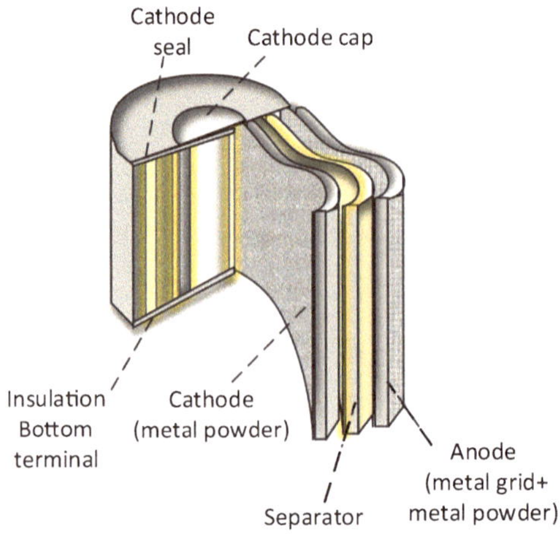

Fig. 1.10 Components of a cylindrical nickel-metal hydride battery cell.

internal pressure increase. The internal pressure increase occurs during the repeated occurrence of overdischarge in which hydrogen evolution takes place at the cathode. The roll design of the nickel-based battery is capable of relatively high electrical power delivery due to higher surface contact between the electrolyte and the electrodes and lower internal resistance. In general, the power yield from the battery increases with an increase in the reaction surface area at the electrodes, a reduction in the collection of current and improvements in the ionic conductivity of the electrolyte.

Overcharging of a Ni-MH battery cell occurs when no nickel hydroxide remains to react with the hydroxide in the electrolyte. Then the oxygen starts to evolve, diffuses across the separator and reaches the anode. There, the oxygen is recombined with the stored hydrogen to create excess water in the potassium hydroxide solution (electrolyte). When the rate at which oxygen evolves from the cathode is higher than the rate at which it recombines, the resulting excess oxygen gas increases the internal pressure of the Ni-MH cell. An effective way to prevent overcharge in a Ni-MH battery cell is to have the anode fabricated with a higher capacity (i.e., more active material) than the cathode. Through this method, effective recombination of the evolved oxygen can be achieved, decelerating the pressure rise inside the cell.

The capacity and lifecycle of a Ni-MH battery cell can be reduced by frequent overcharging. When the pressure inside the cell increases due to the presence of excess oxygen, the temperature also increases within the cell, and this may result in loss of electrolyte in the separator. This is referred to separator dry out, and it diminishes the proper movement of hydrogen between the anode and cathode. In the case of severe overcharging, when there are large amounts of excess oxygen, a safety vent may be used in the positive terminal to ventilate the oxygen gas.

The overdischarging of a Ni-MH battery cell occurs when the active material on the cathode is completely consumed producing hydrogen gas. The metal hydride active material on the anode is capable of absorbing a part of the produced hydrogen gas evolved from the cathode. The hydrogen that is not absorbed by the anode increases the internal pressure of the cell. Then, when the active material on the anode is also completely consumed, the anode starts to absorb oxygen gas, which results in a loss of capacity. By frequent overdischarge of a Ni-MH battery cell, the surfaces of electrodes are fully covered by the hydrogen and oxygen gases, leading to a reduction in storage capacity (as the oxygen gas occupies the hydrogen storage site) and a loss of hydrogen gas as the excess hydrogen is released through a safety vent.

1.3.2.5 Nickel-zinc batteries

In nickel-zinc (Ni-Zn) battery cells, metallic zinc is used as the anode, nickel oxyhydroxide as the cathode and potassium hydroxide solution as the electrolyte. Fig. 1.11 shows the working principle of a nickel-zinc battery cell during charge and discharge. During discharge, at the anode, the metallic zinc is oxidized to zinc (II) hydroxide. During the anodic reaction, two electrons are produced. These are transferred to the cathode through an external circuit. At the cathode, nickel (III)-oxy hydroxide is reduced to nickel (II) hydroxide. In fact, at the cathode, the nickel (III)-oxy hydroxide reacts with water and two electrons (which are produced in the anodic reaction) and form nickel (II) hydroxide and hydroxide ions.

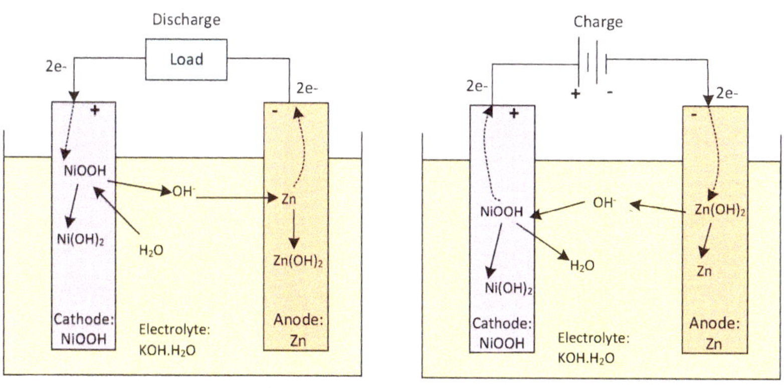

Fig. 1.11 Working principle of nickel-zinc battery cell during discharge and charge.

The redox reactions at the anode and cathode during discharge can be written for Ni-Zn battery cells as follows:

$$\text{Oxidation in anode}: Zn + 2OH^- \rightarrow Zn(OH)_2 + 2e^-; E_a^{\circ} = -1.2\ V \tag{1.30}$$

$$\text{Reduction in cathode}: 2NiOOH + 2H_2O + 2e^- \rightarrow 2Ni(OH)_2 + 2OH^-; E_c^{\circ} = 0.49\ V \tag{1.31}$$

$$\text{Overall reaction}: 2NiOOH + Zn + H_2O \rightarrow ZnO + 2Ni(OH)_2; E_{cell}^{\circ} = 1.7\ V \tag{1.32}$$

The overall cell potential of a Ni-Zn cell is observed to be higher than that of other nickel-based cells. So, this cell type is capable of higher cell voltage delivery. In addition, the zinc element is relatively more abundant than cadmium and metal hydrides, which makes the cost of Ni-Zn cells relatively low. The addition of lithium hydroxide (LiOH) to the electrolyte enhances the performance and lifetime of Ni-Zn battery cells. Ni-Zn battery cells have relatively high energy densities (between 55 and 85 Wh/kg) and power densities (between 140 and 200 W/kg), and low self-discharge rates (below 0.8% per day). Wicking materials are usually used to keep the anode and cathode wet. A safety vent can be used to release the O_2 gas evolved during charging of Ni-Zn battery cells. Ni-Zn battery cells are maintained at high pressure to mitigate the occurrence of deformity during their service lives.

During charging, zinc is precipitated at the anode which results in the formation of dendrites and electrode swelling. These can lead to an internal short circuit. Fig. 1.12 shows dendrite formation from zinc in a Ni-Zn battery cell during repeated charge/discharge cycles. A microporous separator can be used to prevent the formation of zinc dendrites. Nevertheless, further improvements are required to overcome the obstacles associated with zinc plating and dendrites in Ni-Zn battery cells.

1.3.3 Sodium-beta batteries

In sodium-beta battery cells, a solid beta alumina (β-Al_2O_3) material is used as the electrolyte. It is permeable to sodium ions, allowing them to be transported between the anode and the cathode. A sodium electrode is used as the anode. The cathode can be a liquid sulfur (in which case the battery type is sodium-sulfur (Na-S)) or a solid metal-chloride. In Na-S battery cells, a porous graphite can be used in the cathode to improve its electrical conductivity. Solid beta alumina is a ceramic material that has high ionic conductivity at relatively high temperatures (above 300°C). As the melting points of

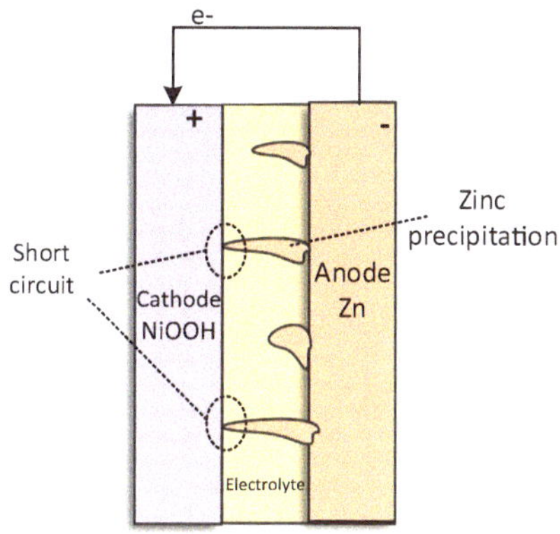

Fig. 1.12 Dendrite formation from zinc in a Ni-Zn battery cell during repeated charge/discharge cycles.

materials used in the anode and the cathode are below 300°C, both the anode and the cathode are in the molten salt state during operation of a sodium-beta battery cell.

This type of battery cell operates effectively at high temperatures, which is an important advantage since it indicates a low sensitivity to ambient temperature. However, the relatively high operating temperatures of such batteries reduce their energy efficiencies and longevities and add more greater safety considerations.

The redox reactions at the anode and cathode of a sodium-sulfur (Na-S) battery cell during discharge at an operating temperature of 350°C can be expressed as follows:

$$\text{Oxidation at anode}: 2\text{Na} \rightarrow 2\text{Na}^+ + 2\text{e} \quad (1.33)$$

$$\text{Reduction at cathode}: x\text{S} + 2\text{Na}^+ + 2\text{e}^- \rightarrow \text{Na}_2\text{S}_x \quad (x = 3-5) \quad (1.34)$$

$$\text{Overall reaction}: x\text{S} + 2\text{Na} \rightarrow \text{Na}_2\text{S}_x \quad (x = 3-5); \quad E^{\circ}_{\text{cell}} = 1.78 - 2.08\ \text{V} \quad (1.35)$$

Fig. 1.13 shows a tubular sodium-nickel chloride battery cell along with the cell structure. The redox reactions at the anode and the cathode at an operating temperature of 300°C for a sodium-beta battery cell which employs nickel chloride as the cathode can be written as:

$$\text{Oxidation at anode}: 2\text{Na} \rightarrow 2\text{Na}^+ + 2\text{e}^- \quad (1.36)$$

$$\text{Reduction at cathode}: \text{NiCl}_2 + 2\text{Na}^+ + 2\text{e}^- \rightarrow \text{Ni} + 2\text{NaCl} \quad (1.37)$$

$$\text{Overall reaction}: \text{NiCl}_2 + 2\text{Na} \rightarrow \text{Ni} + 2\text{NaCl}; E^{\circ}_{\text{cell}} = 2.58\ \text{V} \quad (1.38)$$

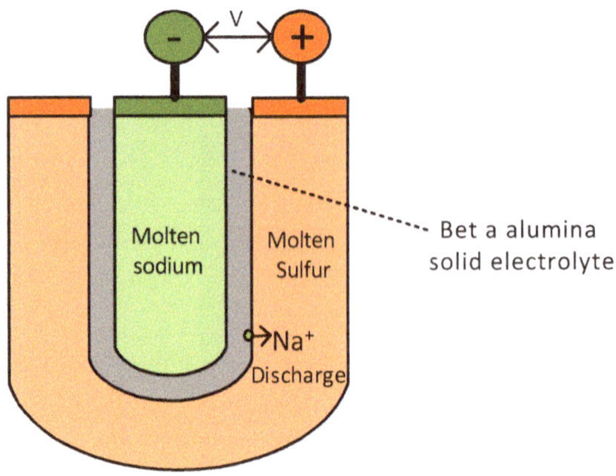

Fig. 1.13 Tubular sodium-nickel chloride battery cell.

During charging, the above redox reactions occur in the reverse direction.

During discharge, the sodium ions are produced through oxidation of metallic sodium at the interface between sodium anode and β-Al_2O_3 electrolyte. Then, the sodium ions are transferred across the electrolyte and react with $NiCl_2$ at the cathode. Finally, $NiCl_2$ is reduced to nickel metal at the cathode.

Sodium beta batteries have relatively high energy densities (five times higher than lead acid batteries), broad energy storage capacities (ranging from kWh to MWh), and long lifecycles (up to 5000 discharge cycles). However, the high temperature operation intensifies corrosion in sodium-sulfur battery cells and mechanical stresses due to thermal expansion within the cells. In addition, the fragility of the electrolyte may result in cracks in the electrolyte structure and subsequently the combination of two liquid cathodes, which can lead to fires and/or explosions. To mitigate this risk, sodium-sulfur batteries should operate above the electrolyte freezing point, which can be achieved by providing an external heat source. By way of comparison, it is pointed out that nickel chloride-based cathodes have higher voltages, higher operating temperatures, a less corrosive nature, safer construction and lower energy densities than the sulfur-based cathode in sodium-beta battery cells.

1.3.4 Lithium ion batteries

In lithium ion batteries, the lithium ion (Li^+) commutes between the anode and the cathode through an electrolyte and porous separator. In the lithium ion battery cells, metal oxides, olivines ($LiMPO_4$ where M = Fe, Mn, Co, Ni), vanadium oxide and lithium-manganese spinels ($LiMn_2O_4$, $LiNi_{0.5}Mn_{1.5}O_4$) are often used as the cathode. Porous carbon, lithium titanium oxide and silicon are often used as the anode. A mixture of organic solvents containing a dissolved lithium salt, like ethylene carbonate ($C_3H_4O_3$), dimethyl carbonate ($C_3H_6O_3$), and propylene carbonate ($C_4H_6O_3$), can be used as the electrolyte in the lithium ion battery.

Fig. 1.14 shows working principle of a typical lithium ion battery. When the lithium ion battery cell is completely discharged, the ions of lithium are in the cathode. During charging, the lithium ions de-intercalate from the cathode and move toward the anode through the electrolyte. Therefore, at the end of the charge the anode is the source of lithium ions, and these start to move back to the cathode during discharge process. Current collectors can be also used at the anode and cathode sides to improve the integrity of the electrodes and adhesion between current collectors and electrodes.

Table 1.2 lists characteristics of various lithium metal battery systems. It is seen that nickel-cobalt-aluminum oxide (NCA)-graphite has the highest specific energy compared to the other types of lithium ion cell chemistries. The operating temperatures of all types of lithium ion batteries are almost the same (between 0°C and 50°C). Operating temperatures less than 20°C and more than 50°C adversely affect the performance of lithium ion batteries due to the acceleration of the degradation mechanisms within the cells.

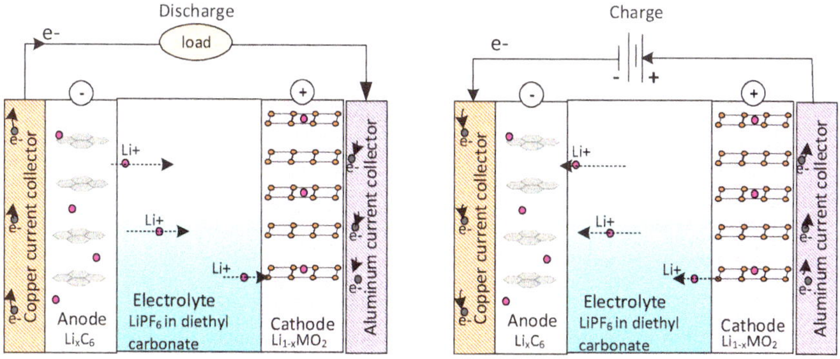

Fig. 1.14 Working principle of a typical lithium ion battery.

Table 1.2 Characteristics of various lithium metal battery systems.

Lithium ion battery chemistry	Cell specific energy (Wh/kg)	Nominal voltage (V)	Cycle life (cycles)	Operation temperature (°C)
NMC-graphite	140–200	3.7	2000+	0–55
NCA-graphite	200–250	3.6	2000+	0–55
LFP-graphite	90–140	3.2	3000+	0–55
LMO-graphite	100–140	3.7	1000–2000	0–55

LFP, lithium iron phosphate; *LMO*, lithium ion manganese oxide; *NCA*, nickel-cobalt-aluminum oxide; *NMC*, nickel-manganese-cobalt-oxide.
Data from Ralon, P., Taylor, M., Ilas, A., Diaz-Bone, H., Kairies, K., 2017. Electricity Storage and Renewables: Costs and Markets to 2030. International Renewable Energy Agency, Abu Dhabi, UAE; Ecker, M., Nieto, N., Käbitz, S., Schmalstieg, J., Blanke, H., Warnecke, A., Sauer, D.U., 2014. Calendar and cycle life study of Li (NiMnCo) O_2-based 18650 lithium-ion batteries. J. Power Sources 248, 839–851; Zubi, G., Dufo-López, R., Carvalho, M., Pasaoglu, G., 2018. The lithium-ion battery: state of the art and future perspectives. Renew. Sustain. Energy Rev. 89, 292–308.

During discharging, the general redox reactions at the anode and cathode for a lithium ion battery using lithium metal oxide ($Li_{(1-x)}MO_2$) as the cathode and graphite (C_6) as the anode can be written as follows:

$$\text{Oxidation at anode}: Li_xC_6 \rightarrow C_6 + xe^- + xLi^+ \quad (1.39)$$

$$\text{Reduction at cathode}: Li_{(1-x)}MO_2 + xe^- + xLi^+ \rightarrow LiMO_2 \quad (1.40)$$

$$\text{Overall reaction}: Li_xC_6 + Li_{(1-x)}MO_2 \rightarrow C_6 + LiMO_2 \quad (1.41)$$

For example, for a lithium cobalt oxide battery cell, the redox reactions at the anode and cathode during discharging can be written as:

$$\text{Oxidation at anode}: LiC_6 \rightarrow C_6 + e^- + Li^+ \quad (1.42)$$

$$\text{Reduction at cathode}: CoO_2 + e^- + Li^+ \rightarrow LiCoO_2 \quad (1.43)$$

$$\text{Overall reaction}: LiC_6 + CoO_2 \rightarrow C_6 + LiCoO_2 \quad (1.44)$$

During charging, the above redox reactions occur in the reverse direction. During discharging, the lithium is oxidized from Li to Li^+ at the lithium graphite anode. The oxidation reaction at the anode also may produce uncharged material that remains at the anode. The produced lithium ions (Li^+) are transferred through the electrolyte to the cathode where they combine with cobalt oxide, leading to the reduction of Co (IV) to Co (III).

Lithium ion battery cells have low self-discharge rates (1.5%–2% per month) and high energy densities (100–265 Wh/kg on a mass basis and

250–670 Wh/L on a volume basis), long lifecycles and low maintenance costs. The lithium ion battery is capable of a voltage delivery of up to 3.6 V, which is three times higher than for Ni-MH and Ni-Cd batteries. Despite of their advantages, lithium ion battery cells tend to overheat during high voltage operation, which can lead to thermal runaway, fire and combustion. Also, lithium ion batteries usually have higher costs than other types of batteries like Ni-Cd cells.

1.3.5 Metal-air batteries

In a metal-air battery cell, oxygen, preferably supplied from the atmosphere, is used as the cathode, a pure metal (e.g., zinc, lithium) for the anode and an aqueous solution (e.g., aqueous alkaline like potassium hydroxide, sodium hydroxide and lithium hydroxide) as the electrolyte. A porous carbon plate is usually used as a support for the air cathode. Various metals can be used for the anode, including iron, zinc, magnesium, lithium, and aluminum. However, only zinc and lithium materials are applicable for rechargeable batteries. During discharging, the oxygen reacts with the pure metal at the anode and forms a metal oxide. During charging, the produced metal oxides are reduced, converting them back to the metal at the anode.

Fig. 1.15 shows a lithium-air battery cell and its working principles. During discharge, lithium ions are generated in the anodic reaction and move toward the cathode through the electrolyte. They react with oxygen from the air to produce lithium peroxide (Li_2O_2) at the cathode.

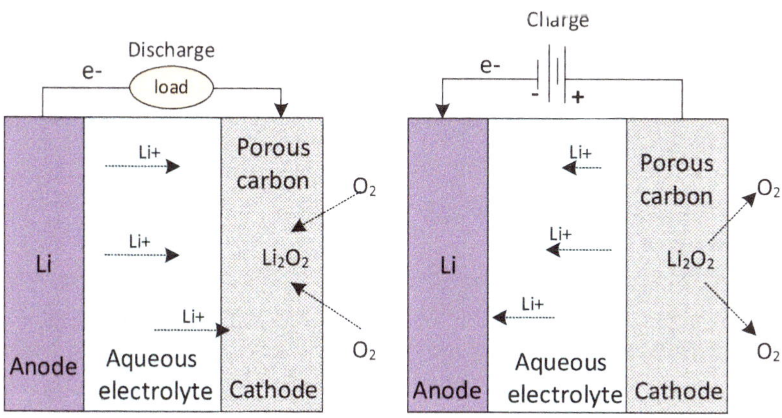

Fig. 1.15 Working principle of lithium-air battery cell.

The redox reactions at the anode and cathode during discharging can be written as:

$$\text{Oxidation in anode}: Li \rightarrow e^- + Li^+ \quad (1.45)$$

$$\text{Reduction in cathode}: 2Li^+ + 2e^- + O_2(g) \rightarrow Li_2O_2(s) \quad (1.46)$$

$$\text{Overall cell reaction}: 2Li + O_2(g) \rightarrow 2Li_2O_2(s) \quad (1.47)$$

Metal-air battery cells have the highest energy densities and lowest energy storage costs compared to any existing battery. The specific energy densities for zinc-air and lithium-air battery cells respectively are 3 and 11 kWh/kg. Despite of their high energy densities, there are notable concerns in the operation of metal-air batteries. These include capacity loss and decline in rate capability due to precipitation of the metal oxide insulating layer on the surface of the anode, and evaporation of the electrolyte and reaction of the metal with external components due to exposure to the atmosphere.

1.4 Application of batteries

Commercialized secondary (rechargeable) battery cells and their working principles were presented earlier. In this section, applications of various battery types are listed and described.

Lead acid batteries: Lead acid batteries are one of the earliest-used batteries and they are still one of the most widely used secondary (rechargeable) electrical storage technologies for small-medium scale storage applications. Lead acid batteries are one of the most popular electrical energy storage types for use in battery-based off-grid and backup systems due to their good economics (i.e., low cost per energy stored) and their balance between battery lifetime, capacity and cost. Also, lead acid batteries are a good option for solar installers. Such batteries are scalable to meet small to large capacity storage needs for solar-based applications like solar photovoltaic systems. Lead acid batteries remain commonly used starter batteries in vehicles based on internal combustion engines.

Nickel-based batteries: Generally, nickel-based batteries are often used in backup power applications (e.g., for supplying emergency power), in hybrid electric vehicle and electric vehicles, and in starting systems for aircraft. Nickel-cadmium batteries are often used when high discharge rates and large capacities are required. Hence, they are commonly used in home power tools, remote control cars, etc., which require high power at high discharge rates. Many solar lights or garden path lights are equipped with nickel-cadmium batteries. Nickel-cadmium batteries can also be used in larger scales as emergency backups for nuclear power plants, aircraft avionic systems and heating ventilation and air conditioning (HVAC) applications.

A higher battery capacity and similar charge rate is observed in nickel-metal hydride batteries. Like nickel-cadmium batteries, nickel-metal hydride batteries are typically used for remote controls and power tools.

Nickel-iron batteries are suitable for use in renewable energy-based systems (e.g., solar energy systems), because they are robust and durable and have long lifetimes and low costs. Recent research has focused on combined batteries and electrolysis for hydrogen and electrical power production, simultaneously (Mulder et al., 2017). As mentioned before, nickel-iron batteries produce hydrogen when fully charged. This produced hydrogen can be used in fuel cell applications for electrical power production.

Nickel-hydrogen batteries have been developed for increasing the capacity and energy density of secondary battery applications in aerospace energy storage. For instance, in Mercury Messenger (NASA, 2004) and Mars Odyssey (Anderson and Coyne, 2002) nickel-hydrogen batteries were initially used.

Sodium-sulfur batteries: Sodium-sulfur batteries have high specific energy densities, which make them attractive as a rechargeable traction battery for automobile applications. However, they have not been successful for commercial use in automobile applications and further research and improvements are still required for large-scale and reliable applications of this battery technology. Sodium-sulfur batteries are currently used in limited numbers for utility load-leveling applications (Santhanagopalan et al., 2014).

Lithium-ion batteries: Lithium-ion batteries are suitable for portable electronic devices such as computers, laptops, cell phones, digital cameras, toys, and power tools, due to their high specific energy, durability and relatively high power operation. In addition, this type of battery is used in extensively in automotive applications like power packs for electric vehicles and hybrid electric vehicles. Also, due to the long lifetimes, high efficiency and high power features of lithium ion batteries, they are currently used in stationary power plants and renewable energy-based systems (e.g., solar and wind).

Numerous applications are being made of lithium-ion batteries. Some of the main applications follow:

- Emergency power backups.
- Electric vehicles and hybrid electric vehicles.

- Lightweight marine performance.
- Solar power storage.
- Alarm systems in remote locations.
- Potable electric devices (laptops, cell phones, etc.).

1.5 Comparison of battery types

In this section, characteristics of various battery types are compared and contrasted.

Fig. 1.16 compares the specific energy density and volumetric energy density for various types of battery cells. It is seen that lithium ion batteries have relatively higher volumetric energy densities and specific energy densities compared to lead acid, nickel cadmium, and nickel metal hydride batteries. These beneficial characteristics are in addition to lithium ion batteries having smaller sizes and lower weights, for a given capacity, as highlighted in Fig. 1.16.

There has been a significant shift toward passenger electric vehicles in transportation. More efforts are required for electrification of light and heavy duty trucks, which are one of the main contributors to carbon dioxide emissions associated with the transportation sector (EPA, 2019). Table 1.3

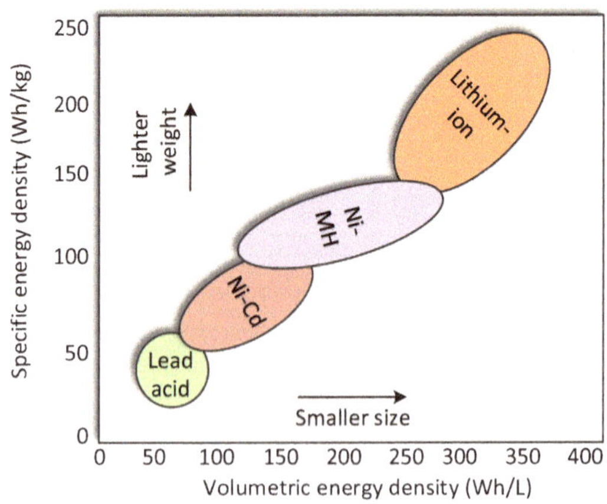

Fig. 1.16 Relation between specific energy density and volumetric energy density for lead acid, nickel cadmium, nickel metal hydride, and lithium ion battery cells. *(Modified from Tarascon, J.M., Armand, M., 2011. Issues and challenges facing rechargeable lithium batteries. In: Dusastre, V. (Ed.), Materials for Sustainable Energy: A Collection of Peer-Reviewed Research and Review Articles From Nature Publishing Group. World Scientific: Nature Pub. Group, Hackensack, NJ, USA, pp. 171–179.)*

Table 1.3 Characteristics and features of commercialized secondary batteries.

Battery type	Lead acid	Nickel-based	Lithium ion
Parameter			
Cost	Cheap	Expensive	Average
Maintenance requirements	High	Low	Moderate
Estimated lifetime (years)	Short	Long	Moderate-long
Depth of discharge	20% for 500 cycles	20% for 2500 cycles	20% for 3000 cycles
Most suitable applications	Emergency lighting, solar power storage and photovoltaic systems, car batteries	Aircraft applications, electric vehicles, power tools, emergency lightening	Electric vehicles, cell phones, laptops, computers
Best charging technique	Constant current-constant voltage	Constant current	Constant current-constant voltage

Data from GREENRHINO, 2021. Beta Batteries, Safety, Storage, Operations, and Maintenance Procedure. http://leadcrystalbatteries.com (Accessed 25 May 2022); Deltech, 2021. Deltech Energy Solution. Available from: https://www.delteconline.co.za/ (Accessed 25 May 2022).

compares the lifetime, cost, depth of discharge (the capacity that is discharged from a fully charged battery divided by battery nominal capacity), application and other characteristics of lithium ion, nickel-based and lead acid batteries. It can be seen that lithium ion batteries have suitable costs, maintenance requirements and lifetimes. Also, nickel-based batteries are the most expensive type, followed by lithium ion and then lead acid batteries.

Much research has been carried out that aims either to improve the performance and cost of existing battery technologies, sometimes to reach preset targets, or to propose new chemistries for the anodes, cathodes, and electrolytes used in batteries. For example, several recent studies have focused on employing new methods to improve the performance of lithium ion batteries (Zhu et al., 2021; Li et al., 2021; Zhu et al., 2020; Reizabal et al., 2020; Shang et al., 2019; Al-Zareer et al., 2019a,b; Malik et al., 2018; Liu et al., 2018a,b; Brown et al., 2018; Wang et al., 2017). Also, studies have been carried out in which performance improvement methods are proposed and discussed for nickel-based batteries (Tang et al., 2022; Shangguan et al., 2020; Li et al., 2017, 2019) and lead acid batteries (Lopes and Stamenkovic, (2020); Wu et al., 2020; Thangarasu et al., 2020).

In addition, various new chemistries have been proposed for batteries and some are under development. The functions, performances capabilities, and technical challenges of these new chemistries are under receiving attention and are under investigation. Examples of newly proposed types of batteries include sodium-iron, liquid-metal and lithium-sulfur batteries. Also, recent studies have been devoted to investigating the operation and performance of new materials for battery cell structures (Yuan et al., 2021; Özdogru et al., 2021; Yeo et al., 2021; Zhang et al., 2021; Wan et al., 2021; Ng et al., 2021).

1.6 Closing remarks

In this chapter, battery cells are introduced and their working principles and characteristics are described, focusing on commercialized secondary (rechargeable) batteries. Applications and characteristics of various battery types are also presented and compared. The abundance of materials used in the structures of batteries is one criterion for determining their costs and availabilities. Further efforts are needed for reducing or minimizing the environmental impacts of batteries and increasing their sustainability. The selection of materials that result in cost-effective and eco-efficient designs of batteries and developing the new technologies applicable for new chemistries with low or minimum environmental impacts are both important factors in advancing battery technology. To achieve this, analyses for material abundance and eco-efficient production processes need to be incorporated in the design of new and existing battery systems.

Study questions

1.1. Define the following terms used in battery systems and explain their differences:
- electrode
- anode
- cathode
- electrolyte
- separator
- current collector

1.2. What is redox reaction? What is the general chemical equation for a redox reaction in an electrochemical system like a battery?

1.3. What group of metals are typically used for the anode and the cathode in a battery cell?

1.4. How can the electrical potential output of a battery system be increased? How can both the electrical potential and current of a battery system be increased, simultaneously?

1.5. Define battery capacity and explain how it can be determined.

1.6. In the below list, connect the electrochemical reactions occurring at the anode and cathode to the corresponding type of battery cell. Note that one battery cell type in the right column is extra and does not correspond to any listed electrochemical reaction.

Overall electrochemical reaction in battery cell	Battery cell type
$Pb\ (s) + PbO_2(s) + 2H_2SO_4(aq) \rightarrow 2PbSO_4(s) + 2H_2O\ (l)$	Nickel cadmium
$2NiOOH + Cd + 2H_2O \rightarrow 2Ni(OH)_2 + Cd(OH)_2$	Sodium nickel chloride
$2NiOOH + Fe + 2H_2O \rightarrow 2Ni(OH)_2 + Fe(OH)_2$	Lead acid
$NiOOH + MH \rightarrow M + Ni(OH)_2$	Nickel iron
$NiCl_2 + 2Na \rightarrow Ni + 2NaCl$	Lithium cobalt oxide
$LiC_6 + CoO_2 \rightarrow C_6 + LiCoO_2$	Nickel metal hydride
	Nickel zinc

1.7. Describe the general applications of lead acid, nickel-based and lithium ion batteries.

1.8. Compare the specific energy density and volumetric energy density for lead acid, nickel cadmium, nickel metal hydride, and lithium ion battery cells.

1.9. Compare lead acid, nickel-based and lithium ion battery cells in terms of lifetime and cost.

1.10. How is the electrical potential of an electrode (cathode or anode) measured through a reference electrode? Explain with an example.

References

Al-Zareer, M., Dincer, I., Rosen, M.A., 2019a. Development and analysis of a new tube based cylindrical battery cooling system with liquid to vapor phase change. Int. J. Refrig. 108, 163–173.

Al-Zareer, M., Dincer, I., Rosen, M.A., 2019b. A novel approach for performance improvement of liquid to vapor based battery cooling systems. Energy Convers. Manage. 187, 191–204.

Anderson, P.M., Coyne, J.W., 2002. A lightweight, high reliability, single battery power system for interplanetary spacecraft. In: Proceedings, IEEE Aerospace Conference. vol. 5, p. 5.

Brown, Z.L., Jurng, S., Nguyen, C.C., Lucht, B.L., 2018. Effect of fluoroethylene carbonate electrolytes on the nanostructure of the solid electrolyte interphase and performance of lithium metal anodes. ACS Appl. Energy Mater. 1 (7), 3057–3062.

EPA, 2019. Sources of Greenhouse Gas Emissions. United States Environmental Protection Agency. Available from: https://www.epa.gov/ghgemissions/sources-greenhouse-gas-emissions. (Accessed 25 May 2022).

IEA, 2021. Energy Storage. International Energy Agency, Paris. Available from: https://www.iea.org/reports/energy-storage. (Accessed 25 May 2022).

Li, W., Zhang, F., Xiang, X., Zhang, X., 2017. High-efficiency Na-storage performance of a nickel-based ferricyanide cathode in high-concentration electrolytes for aqueous sodium-ion batteries. ChemElectroChem 4 (11), 2870–2876.

Li, J., Zhang, H., Wu, C., Cai, X., Wang, M., Li, Q., Chang, Z., Shangguan, E., 2019. Enhancing the high-temperature and high-rate properties of nickel hydroxide electrode for nickel-based secondary batteries by using nanoscale $Ca(OH)_2$ and γ-CoOOH. J. Electrochem. Soc. 166 (10), 1836.

Larcher, D, Tarascon, JM, 2015. Towards greener and more sustainable batteries for electrical energy storage. Nat. Chem. 7 (1), 19–29.

Li, L., Shan, Y., Wang, F., Chen, X., Zhao, Y., Zhou, D., Wang, H., Cui, W., 2021. Improving fast and safe transfer of lithium ions in solid-state lithium batteries by porosity and channel structure of polymer electrolyte. ACS Appl. Mater. Interfaces 13 (41), 48525–48535.

Liu, K., Liu, Y., Lin, D., Pei, A., Cui, Y., 2018a. Materials for lithium-ion battery safety. Sci. Adv. 4 (6), 9820.

Liu, W., Li, X., Xiong, D., Hao, Y., Li, J., Kou, H., Yan, B., Li, D., Lu, S., Koo, A., Adair, K., 2018b. Significantly improving cycling performance of cathodes in lithium ion batteries: the effect of Al_2O_3 and $LiAlO_2$ coatings on $LiNi_{0.6}Co_{0.2}Mn_{0.2}O_2$. Nano Energy 44, 111–120.

Lopes, PP, Stamenkovic, VR, 2020. Past, present, and future of lead–acid batteries. Science 369 (6506), 923–924.

Malik, M., Mathew, M., Dincer, I., Rosen, M.A., McGrory, J., Fowler, M., 2018. Experimental investigation and thermal modelling of a series connected $LiFePO_4$ battery pack. Int. J. Therm. Sci. 132, 466–477.

Mulder, F.M., Weninger, B.M.H., Middelkoop, J., Ooms, F.G.B., Schreuders, H., 2017. Efficient electricity storage with a battolyser, an integrated Ni–Fe battery and electrolyser. Energy Environ. Sci. 10 (3), 756–764.

NASA, 1997. Lesson Number 568. The National Aeronautics and Space Administration. Available from: https://llis.nasa.gov/lesson/568. (Accessed 25 May 2022).

NASA, 2004. Messenger, NASA's Mission to Mercury. The National Aeronautics and Space Administration. Available from: www.nasa.gov/pdf/168019main_MESSENGER_71504_PressKit.pdf. (Accessed 25 May 2022).

Ng, S.F., Lau, M.Y.L., Ong, W.J., 2021. Lithium–sulfur battery cathode design: tailoring metal-based nanostructures for robust polysulfide adsorption and catalytic conversion. Adv. Mater. 33, 2008654.

Özdogru, B., Dykes, H., Gregory, D., Saurel, D., Murugesan, V., Casas-Cabanas, M., Çapraz, Ö.Ö., 2021. Elucidating cycling rate-dependent electrochemical strains in sodium iron phosphate cathodes for Na-ion batteries. J. Power Sources 507, 230297.

Reizabal, A., Gonçalves, R., Fidalgo-Marijuan, A., Costa, C.M., Pérez, L., Vilas, J.L., Lanceros-Mendez, S., 2020. Tailoring silk fibroin separator membranes pore size for improving performance of lithium ion batteries. J. Membr. Sci. 598, 117678.

Santhanagopalan, S., Smith, K., Neubauer, J., Kim, G.H., Keyser, M., Pesaran, A., 2014. Design and Analysis of Large Lithium-Ion Battery Systems. Artech House.

Shang, Z., Qi, H., Liu, X., Ouyang, C., Wang, Y., 2019. Structural optimization of lithium-ion battery for improving thermal performance based on a liquid cooling system. Int. J. Heat Mass Transf. 130, 33–41.

Shangguan, E., Zhang, H., Wu, C., Cai, X., Wang, Z., Wang, M., Li, L., Wang, G., Li, Q., Li, J., 2020. CoAl-layered double hydroxide nanosheets-coated spherical nickel hydroxide cathode materials with enhanced high-rate and cycling performance for alkaline nickel-based secondary batteries. Electrochim. Acta 330, 135198.

Shukla, A.K., Venugopalan, S., Hariprakash, B., 2001. Nickel-based rechargeable batteries. J. Power Sources 100 (1–2), 125–148.

Tang, Y., Guo, W., Zou, R., 2022. Nickel-based bimetallic battery-type materials for asymmetric supercapacitors. Coord. Chem. Rev. 451, 214242.

Thangarasu, S., Palanisamy, G., Roh, S.H., Jung, H.Y., 2020. Nanoconfinement and interfacial effect of Pb nanoparticles into nanoporous carbon as a longer-lifespan negative electrode material for hybrid lead carbon battery. ACS Sustain. Chem. Eng. 8 (23), 8868–8879.

Wan, H., Liu, S., Deng, T., Xu, J., Zhang, J., He, X., Ji, X., Yao, X., Wang, C., 2021. Bifunctional interphase-enabled $Li_{10}GeP_2S_{12}$ electrolytes for lithium–sulfur battery. ACS Energy Lett. 6 (3), 862–868.

Wang, M., Zhang, R., Gong, Y., Su, Y., Xiang, D., Chen, L., Chen, Y., Luo, M., Chu, M., 2017. Improved electrochemical performance of the $LiNi_{0.8}Co_{0.1}Mn_{0.1}O_2$ material with lithium-ion conductor coating for lithium-ion batteries. Solid State Ionics 312, 53–60.

Wu, Z., Liu, Y., Deng, C., Zhao, H., Zhao, R., Chen, H., 2020. The critical role of boric acid as electrolyte additive on the electrochemical performance of lead-acid battery. J. Energy Storage 27, 101076.

Yeo, J.S., Yoo, E., Im, C.N., Cho, J.H., 2021. Enhanced electrochemical properties of lithium-tin liquid metal battery via the introduction of bismuth cathode material. Electrochim. Acta 389, 138697.

Yuan, X., Ma, F., Chen, X., Sun, R., Chen, Y., Fu, L., Zhu, Y., Liu, L., Yu, F., Wang, J., Wu, Y., 2021. Aqueous zinc–sodium hybrid battery based on high crystallinity sodium–iron hexacyanoferrate. Mater. Today Energy 20, 100660.

Zhang, S., Liu, Y., Fan, Q., Zhang, C., Zhou, T., Kalantar-Zadeh, K., Guo, Z., 2021. Liquid metal batteries for future energy storage. Energy Environ. Sci. 14, 4177–4202.
Zhu, S., Hu, C., Xu, Y., Jin, Y., Shui, J., 2020. Performance improvement of lithium-ion battery by pulse current. J. Energy Chem. 46, 208–214.
Zhu, W., Kierzek, K., Wang, S., Li, S., Holze, R., Chen, X., 2021. Improved performance in lithium ion battery of CNT-Fe_3O_4 graphene induced by three-dimensional structured construction. Colloids Surf. A Physicochem. Eng. Asp. 612, 126014.

CHAPTER 2

Thermodynamics of batteries

Chapter objectives

- Define enthalpy of reaction and Gibbs function of reaction for a battery system.
- Derive Nernst equation based on the first and second laws of thermodynamics for a generalized battery system.
- Develop a general energy balance for a battery system accounting for electrical potential losses (i.e., overpotentials) in the electrochemical reactions.
- Determine the heat generation rate within a battery.

Nomenclature

Symbols

A	area (m^2)
$\bar{a}$	mean diameter of ions (nm)
A_{DH}	Debye-Hückel constant ($(kg/mol)^{0.5}$)
b	molality (moles of solute per unit mass of solvent) (kmol/kg)
B_{DH}	Debye-Hückel parameter ($(kg/mol)^{0.5}$)/nm)
B	temperature coefficient in expression for equilibrium potential (V/K)
c	molar concentration (mol/cm^3)
c_p	specific heat capacity at constant pressure (kJ/kg K)
E	energy (kJ); potential (V)
F	Faraday's constant (coulomb/mole); view factor
$\bar{g}$	molal Gibbs free energy (kJ/kmol)
G	Gibbs free energy (kJ)
ΔG	Gibbs free energy change of reaction (kJ)
$\overline{h}_f$	molal enthalpy of formation (kJ/kmol)
h_0	heat transfer coefficient (kW/m^2 K)
$\overline{\Delta h}_{sen}$	molal sensible enthalpy (kJ/kmol)
H	enthalpy (kJ)
ΔH_{rxn}	enthalpy of reaction (kJ)
I	cell current (C/s, A)
I_r	partial current of electrode reaction r (A)
I_e	molar ionic strength of electrolyte (mol/kg)
j	current density (A/m^2)
m	mass (kg)
n_e	number of transferred electrons in electrode reaction
$\dot{N}$	mole flow rate (kmol/s)

Battery Technology
https://doi.org/10.1016/B978-0-443-18862-6.00009-4

P pressure (Pa)
$\dot{Q}$ heat transfer rate (kW)
R universal gas constant (8.3143 J/mol K)
$\bar{s}$ molal enthalpy (kJ/kmol K)
S entropy (kJ/K)
$\dot{S}_{gen}$ entropy generation rate (kW/K)
T temperature (°C, K)
T_a ambient temperature (°C, K)
t time (s)
V cell voltage (V)
$\dot{W}$ work exchange rate (kW)
z charge number

Greek letters

α activity
ε surface emissivity
μ chemical/electrochemical potential (kJ/kmol)
γ activity coefficient
$\gamma_{\pm}$ mean molal activity coefficient
ν stoichiometry coefficient

Subscripts

ave average
cv control volume
eq equilibrium
hr half reaction
or overall reaction
P product
R reactant
r reaction
ref reference
rev reversible
rxn reaction
sen sensible

Superscripts

Θ property expressing secondary reference state
$^{\circ}$ property expressing standard state

2.1 Introduction

In this chapter, fundamental concepts important for understanding the thermodynamic and electrochemical characteristics of battery systems are described. This includes the definition of enthalpy of formation, absolute enthalpy, and enthalpy of reaction. Also, the Gibbs function of reaction along with the Nernst equation is derived based on the first and second laws of thermodynamics for a generalized battery system that operates reversibly. Then a general energy balance is developed for a battery system accounting for overpotentials due to irreversibilities in the electrochemical reactions. This formulation is useful for determination of the heat generation rate within a battery and consequently for designing thermal management systems to control battery temperatures.

2.2 Thermodynamics and potential of batteries

In the thermodynamic assessment of an electrochemical system, the absolute molal enthalpy needs to be determined. This thermodynamic property varies with temperature T and pressure P. The absolute molal enthalpy (i.e., enthalpy on a unit mole basis) includes the chemical energy and sensible thermal energy. The chemical energy, known precisely as the molal enthalpy of formation $\overline{h}_f$, denotes the molar enthalpy of a substance due to its chemical composition, while the specific sensible thermal energy $\overline{\Delta h}_{sen}$ expresses the enthalpy of a species with reference to a specified state. Thus, the absolute molal enthalpy $\overline{h}$ can be written as:

$$\overline{h}(T, P) = \overline{h}_f(T_{ref}, P_{ref}) + \overline{\Delta h}_{sen}(T, P) \tag{2.1}$$

In the term containing the enthalpy of formation, the reference temperature T_{ref} and pressure P_{ref} need to be chosen based on the reference conditions. The standard temperature (25°C) and pressure (1 atm) are usually chosen as the reference state at which many of the thermodynamic parameters for various substances are tabulated. Thus, the absolute molal enthalpy can be written as:

$$\overline{h}(T, P) = \overline{h_f}(T^\circ, P^\circ) + \left(\overline{h}_{sen} - \overline{h}^\circ_{sen}\right) \tag{2.2}$$

where T° and P° are the standard reference temperature and pressure which are equal to 25°C and 1 atm, respectively. The term $\overline{h}^\circ_{sen}$ is the standard molal sensible enthalpy at the reference state of 25°C and 1 atm. The molal enthalpy of formation $\overline{h}_f$ is generally determined either from a thermochemical analysis or experimental approaches. Note that the enthalpy of formation for an element in its natural state is always zero, while the enthalpy of formation for a substance comprises of more than one element has a nonzero value. For instance, the enthalpy of formation of oxygen gas (O_2) or solid carbon in the form of graphite (C) are always zero, while the enthalpy of formation of carbon dioxide (CO_2) is nonzero. Specifically, the enthalpy of formation of carbon dioxide is −393.5 kJ per mole of CO_2. The negative sign for the enthalpy of formation indicates the amount of heat released while a positive sign indicates the amount of heat gained during the formation of a compound.

At steady state conditions, the change in energy of a system during a chemical reaction, where there is no work interactions and no change in kinetic and potential energies, is equal to the change in chemical composition (i.e., enthalpy of formation) and change in state (i.e., sensible thermal energy) of system. That is:

$$\Delta E_{sys} = Q = \Delta H_f - \Delta H_{sen} \tag{2.3}$$

where ΔE_{sys} is the energy change of system which is equal to the heat Q absorbed or released during the chemical reaction. When the products leaves the reaction chamber at the inlet state of reactants, the sensible thermal energy term, ΔH_{sen}, becomes zero and at this condition the energy change during the chemical reaction is only due to the change in the chemical composition of the system. This property is called enthalpy of reaction.

The enthalpy of reaction ΔH_{rxn} can be defined as the total thermal energy liberated or absorbed during a complete chemical reaction. This term can be determined through the difference between the enthalpy of products at a specified state and the enthalpy of reactants at the same state for a complete chemical reaction. Consider the following general chemical reaction:

$$\underbrace{aA + bB + \cdots}_{\text{Reactants(R)}} \rightarrow \underbrace{xX + yY + \cdots}_{\text{Products(P)}} \tag{2.4}$$

For an electrochemical reaction in which oxidation and reduction occur, the ions and electrons also contributed in Eq. (2.4) as the reactants and products (i.e., Oxidizing agent(s) + $n_e e^-$ → Reducing agent(s)).

The general expression for the enthalpy of reaction for any chemical/electrochemical reaction can be expressed as

$$\Delta H_{rxn} = Q = \sum_{P} \nu_P \bar{h}_P - \sum_{R} \nu_R \bar{h}_R \tag{2.5}$$

where ν_P and ν_R respectively are the stoichiometry coefficients of each product P ($P = x, y, \ldots$) and reactant R ($R = a, b, \ldots$) which are equivalent to the corresponding number of moles products and reactants. As mentioned before, to determine the enthalpy of reaction for an open system (i.e., control volume), the pressure and temperature of inlet reactants and outlet products need to be identical, while the pressure and temperature at the initial state (before the reaction starts) and final state (when the reaction ends) need to be equal for a closed system. The negative sign for enthalpy of reaction indicates the release of thermal energy during the chemical reaction. In such a case, the reaction is classified as exothermic, while a positive enthalpy of reaction indicates a gain of thermal energy by the system. Then, the reaction is classified as endothermic.

The Gibbs free energy is another important thermodynamic property in an electrochemical system. The Gibbs free energy can be utilized to determine the maximum reversible work done by a thermodynamic system. The Gibbs free energy on a mole basis $\bar{g}$ can be defined as

$$\bar{g} = \bar{h} - T\bar{s} \tag{2.6}$$

where $\bar{h}$, T, and $\bar{s}$ respectively denote molal enthalpy, absolute temperature (K) and molal entropy. As for the enthalpy of reaction, the Gibbs free energy change of a chemical reaction can be defined as the difference between the Gibbs free energy of products at a specified state and the Gibbs free energy of reactants at the same state. That is:

$$\Delta G_{rxn} = \sum_{P} \nu_P \bar{g}_P - \sum_{R} \nu_R \bar{g}_R = \sum_{P} \nu_P \left(\bar{h}_P - T\bar{s}_P\right) - \sum_{R} \nu_R \left(\bar{h}_R - T\bar{s}_R\right) \tag{2.7}$$

where the stoichiometric coefficient ν for each product P and reactant R. The stoichiometric coefficients are equivalent to the corresponding number of moles of products and reactants. In the discharge operation of a battery, the chemical energy content of the chemical compounds is directly converted to electrical work.

2.3 Reversible cell potential

Fig. 2.1 shows a generalized battery system in which the temperature of the cell varies with time and location due to electrochemical reactions, mixing effect, phase change, etc. Consider a discharge/charge operation of battery as occurring in a *control volume* in which the overall cell reaction (i.e., one pair of electrode reactions) take place. Assuming a uniform temperature throughout the battery (i.e., $T_{\text{cell}} = f(t)$) and negligible changes in kinetic and potential energies, the first and second laws of thermodynamics for such a battery can be written as (Dincer and Rosen, 2021; Rosen and Farsi, 2022):

$$\frac{dE_{\text{cv}}}{dt} = -\dot{Q} - \dot{W} + \sum_R \dot{N}_R \overline{h}_R - \sum_P \acute{N}_P \overline{h}_P \tag{2.8}$$

$$\frac{dS_{\text{cv}}}{dt} = -\frac{\dot{Q}}{T} + \sum_R \dot{N}_R \overline{s}_R - \sum_P \acute{N}_P \overline{s}_P + \dot{S}_{\text{gen}} \tag{2.9}$$

Here, dE_{cv}/dt represents the change in the energy of battery with time, $\acute{N}_P$ and $\overline{h}_P$ are the mole flow rate and molal enthalpy of each product, $\acute{N}_R$ and $\overline{h}_R$ are the mole flow rate and molal enthalpy of each reactant. The term $\dot{Q}$ is the heat transfer rate, which embodies all heat transfer rates with the environment and other reservoirs (i.e., $\dot{Q} = \sum \dot{Q}_l$). The term $\dot{W}$ is a general representation for the work exchange rate[l] across the control volume. Generally, for a heat exchange process between the control volume and a

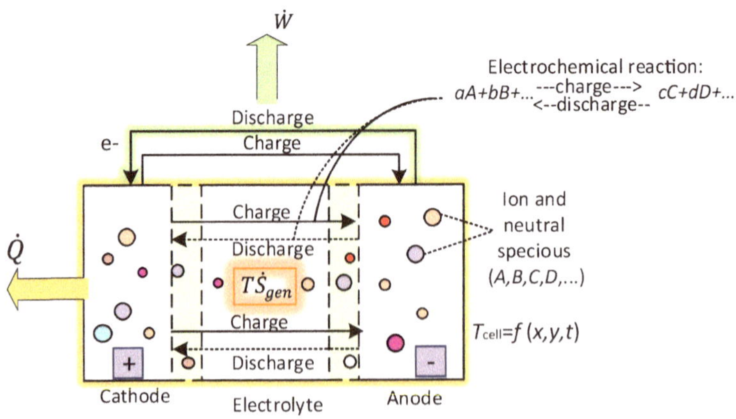

Fig. 2.1 General battery system with energy interactions and electrochemical reactions. The cell temperature T_{cell} is a function of location (x,y) and time t. The electrochemical reactions during charging are reversed during the discharge operation.

reservoir or the environment, heat loss from the control volume is defined as negative and heat gain as positive. For work exchanges, the work done on the control volume is defined as negative and work done by the control volume as positive. In Eq. (2.9), dS_{cv}/dt denotes the change in entropy of the control volume with time, T is the absolute temperature at which the heat transfer process take place, $\bar{s}_P$ and $\bar{s}_R$ are the molal entropies of each product and reactant respectively, and $\dot{S}_{gen}$ is the entropy generation rate within the control volume.

The following assumptions are invoked for the battery system to aid the modeling and analysis:

- All reactants and products are at the same temperature and pressure.
- The battery is in contact with a cooling system that maintains the battery temperature at a specific value (i.e., $dT_{cell}/dt=0$) (see Fig. 2.2).
- The battery operates at the steady state conditions (i.e., $dE_{cv}/dt=0$ and $dS_{cv}/dt=0$).

The terms on the left side of Eqs. (2.8) and (2.9) are zero, because at steady state conditions there is no change associated with the energy and entropy of the system. Thus Eqs. (2.8) and (2.9) can be rewritten respectively as:

$$\dot{W} = -\dot{Q} + \sum_{R} \dot{N}_R \bar{h}_R - \sum_{P} \dot{N}_P \bar{h}_P \tag{2.10}$$

$$\dot{Q} = T\left(\sum_{R} \dot{N}_R \bar{s}_R - \sum_{P} \dot{N}_P \bar{s}_P \right) + T\dot{S}_{gen} \tag{2.11}$$

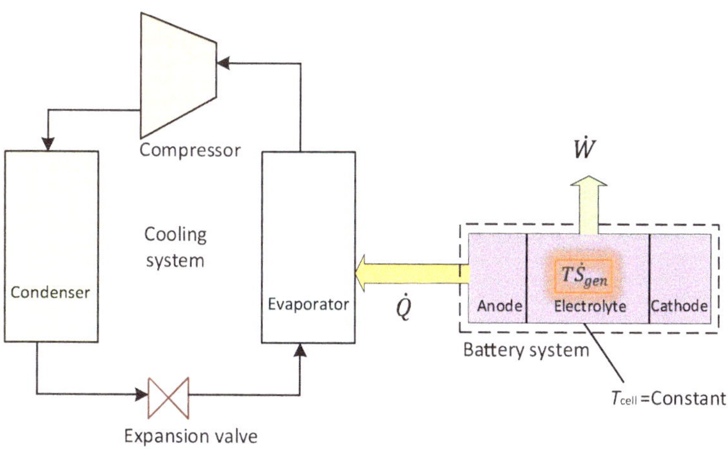

Fig. 2.2 Use of a cooling system to maintain constant temperature of a battery system.

Substituting $\dot{Q}$ from Eq. (2.10) into Eq. (2.11) yields

$$\dot{W} = \left(\sum_R \dot{N}_R \bar{h}_R - \sum_P \dot{N}_P \bar{h}_P\right) - T\left(\sum_R \dot{N}_R \bar{s}_R - \sum_P \dot{N}_P \bar{s}_P\right) - T\dot{S}_{gen} \tag{2.12}$$

Recalling the definition of Gibbs free energy (i.e., $g = h - Ts$), Eq. (2.12) can be expressed in terms of the Gibbs function as follows:

$$\dot{W} = -\left(\sum_P \dot{N}_P \bar{g}_P - \sum_R \dot{N}_R \bar{g}_R\right) - T\dot{S}_{gen} \tag{2.13}$$

where the term in parenthesis denotes the Gibbs free energy change rate during the chemical reaction. In a real system, the entropy generation rate term (i.e., $\dot{S}_{gen}$) is always positive meaning that the entropy of a real system during its operation always increases due to the irreversibilities and losses. In an ideal system, the entropy generation rate is zero, meaning that are is no irreversibilities or losses in the system. Such system is called *reversible*. For a system operating under reversible conditions, the produced work rate is a maximum, and can be determined as:

$$\dot{W}_{rev} = \dot{W}_{max} = -\left(\sum_P \dot{N}_P \bar{g}_P - \sum_R \dot{N}_R \bar{g}_R\right) \tag{2.14}$$

Therefore, for an ideal battery system operating at reversible conditions, the maximum work rate that can be generated during discharge mode is determined by Eq. (2.14). Also, Eq. (2.14) can be expressed in terms of per mole of reducing agent in the chemical reaction occurring in the battery. If the mole flow rate of reducing agent (RA) is $\dot{N}_{R=RA}$, the work rate and reversible work rate produced per mole of reducing agent respectively are given by:

$$\dot{W}/\dot{N}_{R=RA} = w = -\frac{1}{\dot{N}_{R=RA}}\left(\sum_P \dot{N}_P \bar{g}_P - \sum_R \dot{N}_R \bar{g}_R\right) - \frac{1}{\dot{N}_{R=RA}} T\dot{S}_{gen} \tag{2.15}$$

$$\dot{W}_{rev}/\dot{N}_{R=RA} = w_{rev} = -\frac{1}{\dot{N}_{R=RA}}\left(\sum_P \dot{N}_P \bar{g}_P - \sum_R \dot{N}_R \bar{g}_R\right) \tag{2.16}$$

Also, the potential of a battery cell at the reversible condition is a maximum and referred to as the *reversible cell potential* or *equilibrium potential.*

Also, the potential of a cell is a maximum when no current flows through the cell; for instance, when a high impedance (resistance) potentiostat is placed between the electrodes and no current can pass through the cell. This potential is called the *open circuit potential.* The open circuit potential of a battery cell can be measured experimentally and is usually lower than the reversible cell potential due to diffusion irreversibilities. However, the difference between the open circuit cell potential and reversible cell potential is usually insignificant and can be ignored in the analysis.

During the discharge mode of a battery cell, the electromotive force due to the potential difference between the electrodes moves charge in the internal circuits from the negative electrode (anode) to the positive electrode (cathode). During the charge mode, an external electrical potential energy is applied to the cell to move charge from cathode to anode. For a real battery the work generated during discharge mode is a function of cell potential E and charge C as follows:

$$W = EC \tag{2.17}$$

Here, the charge C can be written in terms of the number of transferred electrons in reaction n_e and Faraday's constant F. Thus, Eq. (2.17) becomes:

$$W = EC = E(n_e F) \tag{2.18}$$

As pointed out before, for reversible operation of a battery cell, the generated work is a maximum, and the corresponding cell potential is also a maximum and equal to the equilibrium potential E_{eq} (or reversible potential). So, Eq. (2.18) for this condition can be written as:

$$W_{\text{rev}} = -\Delta G = E_{\text{eq}}(n_e F) \tag{2.19}$$

Therefore, the voltage that is lost due to irreversibilities (i.e., entropy generation) can be deducted from the reversible potential of cell:

$$E = E_{\text{eq}} - \frac{T\dot{S}_{\text{gen}}}{n_e F} \tag{2.20}$$

For a multicomponent/multiphase system like a battery in which electrochemical reactions occur, the Gibbs free energy is related to the electrochemical/chemical potential and molar content of each species i in each phase j, as follows:

$$G = \sum_j \sum_i \nu_i^j \mu_i^j \tag{2.21}$$

where ν_i^j is the stoichiometry coefficient (equivalent to molar amount) of neutral or charged species i in phase j, and μ_i^j is the electrochemical potential of charged species i (or chemical potential of neutral species i) in phase j. Note that each phase is electrically neutral even though all ionic species are not present in each phase. For example, in an electrolyte, the number of anions and cations needs to be equal to satisfy the electrically neutral condition (i.e., $\sum c_i z_i = 0$, where c_i and z_i respectively are molar concentration and charge number of species i). These constraints can be used in the determination of species concentrations. Furthermore, at equilibria between the phases, the electrochemical potentials of any ions are identical for those phases. For instance, in an electrochemical cell where there are two equilibrium phases of electrolyte (ϵ) and solid salt (φ), the electrochemical potential of ion i in both phases are identical (i.e., $\mu_i^\epsilon = \mu_i^\varphi$). One of the approaches for calculation of phase equilibrium is the Gibbs free energy minimization method along with the stoichiometry and electric charge constraints required to determine the equilibrium concentrations of species in an electrolytic cell.

The electrochemical potential used in Eq. (2.21) is related to the electrochemical potential at the secondary reference state ($\mu^\Theta{}_i$) and activity of species i (α_i) as follows:

$$\mu_i = \mu^\Theta{}_i + RT \ln \alpha_i \tag{2.22}$$

where T is the absolute temperature and R is universal gas constant ($R = 8.314\,\mathrm{J\,K^{-1}\,mol^{-1}}$). The parameter α_i is the activity of species i which can be calculated based on the species molality b_i, which represents the moles of solute per unit mass of solvent, and the activity coefficient γ_i as follows:

$$\alpha_i = b_i \gamma_i \tag{2.23}$$

In a battery or any electrochemical cell where nonideal (real) mixtures are present, the *activity coefficient* needs to be defined to account for the deviation from the ideal conditions due to the ionic strength and interaction of a species with all other ions in the electrolyte. For the ideal state of a mixture, the activity coefficient is defined to be unity and this state is called the *secondary reference state*. In other words, an infinitely dilute solution (i.e., concentration of all charged ions and neutral solutes approaches zero) can be considered as the secondary reference state at which the activity of all charged ions and neutral solutes approaches unity.

Alternatively, in many reports in the literature the general term *standard state* is used to define the ideal solute, much like using this term to define the standard state of other phases (e.g., for the gaseous phase the standard state is defined as a pure substance obeying the ideal gas equation at standard pressure (1 bar)). Thermodynamic properties that are measured at the standard state are known as standard thermodynamic properties. Note that different standard states—or secondary reference states—may conventionally defined for gaseous, liquid and solid mixtures. So, it is important to understand what conventions are employed in preparation of standard thermodynamic properties tables before using them to describe solutions. For instance, the SUPCRT92 database program (Johnson et al., 1992) prepared the standard thermodynamic properties of ions based on a standard condition of temperature 25°C, a pressure of 1 atm and a unit concentration of 1 mol/L of solution, while Newman and Thomas-Alyea (2012) defined the infinite dilute solution (i.e., molality of all species approach zero) as the ideal condition for a solution.

The effect of activity coefficients on the thermodynamic characteristics of an electrochemical cell (such as Gibbs free energy, entropy, and enthalpy) should be considered in the thermodynamic model of a battery cell. Thus, the thermodynamic properties of Gibbs free energy, enthalpy, and entropy in nonideal condition can be expressed based on the standard thermodynamic properties as:

$$\bar{g}_i = \bar{g}_i^\circ + RT\ \ln(\alpha_i) \tag{2.24}$$

$$\bar{h}_i = \bar{h}_i^\circ - RT^2 \frac{\partial \ln(\alpha_i)}{\partial T} \tag{2.25}$$

$$\bar{s}_i = \frac{1}{T}\left(\bar{h}_i - \bar{g}_i\right) \tag{2.26}$$

where $\bar{g}_i$, $\bar{h}_i$, and $\bar{s}_i$ are specific molar Gibbs free of energy, specific molar enthalpy, and specific molar entropy of species *i*. The terms $\bar{g}_i^\circ$, $\bar{h}_i^\circ$, and $\bar{s}_i^\circ$ are specific molar Gibbs free energy, enthalpy, and entropy of species *i* at the standard state. It is evident from that $\bar{g}_i^\circ$ in Eq. (2.24) is equivalent to $\mu^\Theta{}_i$ in Eq. (2.22) specifying the Gibbs free energy at the standard state or secondary reference state.

Several empirical correlations have been reported for the activity coefficient based on the concentration of solution. One well-known model for activity coefficient is the Debye-Hückel theory which is the basis for the empirical correlation of activity coefficients for several electrolytes over a range of concentrations. The activity coefficient for a dilute electrolyte solution can be expressed as follows (Newman and Thomas-Alyea, 2012; Zamfirescu et al., 2017; Farsi et al., 2020):

$$\ln \gamma_i = -\frac{1}{1 + B_{DH}\bar{a}\, I_e^{0.5}} \left(A_{DH} z_i^2\, I_e^{0.5} \right) \tag{2.27}$$

where the parameters B_{DH} and A_{DH} are constants defined in the Debye-Hückel model for activity coefficient in an infinitely dilute electrolyte. Their values can be determined through techniques like concentration-cell and melting point expression experiments for an electrolyte solution. The terms z_i and $\bar{a}$ respectively are charge number of species i and mean diameter of ions. The term I_e is the molar ionic strength of the electrolyte, which is related to the concentration of species i (c_i) (i.e., molarity) and the charge number of species i (z_i) as:

$$I_e = \frac{1}{2}\left(\sum_i c_i {z_i}^2 \right) \tag{2.28}$$

Also, Debye and Hückel developed the following equation for mean activity coefficient of a single-solvent electrolyte A (i.e., $i = A$) that dissociates into cations of charge number z^+ and anions of charge number z^- (Newman and Thomas-Alyea, 2012):

$$\ln \gamma_A = \ln \gamma_\pm = -\frac{1}{1 + Ba\, I_e^{0.5}} \left(A_{DH} z^+ z^-\, I_e^{0.5} \right) \tag{2.29}$$

where $\ln \gamma_\pm$ is the mean activity coefficient on a mole-scale which can be measured and tabulated for solutions of single electrolytes.

For a highly concentrated electrolyte solution, additional terms are added to the Debye-Hückel equation to account for effects that are neglected in the Debye-Hückel theory (Newman and Thomas-Alyea, 2012). That is,

$$\ln \gamma_\pm = -\frac{1}{1 + Ba\, I_e^{0.5}} \left(A_{DH} z^+ z^-\, I_e^{0.5} \right) + B_2 b + B_3 b^{\frac{3}{2}} + B_4 b^2 + \ldots + B_n b^{\frac{n}{2}} + \ldots \tag{2.30}$$

Here, b is the molality of species A. Also, the values for $B_{k=2,\,3,\,\ldots}$ can be estimated empirically by fitting Eq. (2.30) to activity coefficient data.

It is known that several simultaneous electrochemical reactions take place in the electrode of a battery system. It is assumed that species i is always in phase m during the cell operation which results in having a specific secondary reference state and any precipitation of that species in the electrolyte is accounted for separately. Also, the sum over phase j becomes redundant, because species i (i.e., summed over, as well) determines phase j. Then, the Gibbs free energy relations in Eqs. (2.22) and (2.24) can be substituted into Eq. (2.19) to yield the following equation for the energy change in a half-reaction under reversible operation of the electrode:

$$W_{\mathrm{rev,hr}} = -\Delta G_{\mathrm{hr}} = -\Delta G_{\mathrm{hr}}{}^{\Theta} + RT\ \ln\left(\frac{\prod (b_i^{\nu_i}\gamma_i^{\nu_i})_R}{\prod (b_i^{\nu_i}\gamma_i^{\nu_i})_P}\right) \tag{2.31}$$

where $\Delta G_{\mathrm{hr}}{}^{\Theta}$ is the standard Gibbs free energy change for the half-reaction (hr) at the anode or the cathodes, at which reactants (R) are converted to products (P).

The energy change for an *overall reaction* (or) of a full cell including half-reactions at the anode and cathodes is given by the change in Gibbs free energy for each half-reaction:

$$W_{\mathrm{rev,or}} = -\Delta G_{\mathrm{ov}} = \Delta G_{\mathrm{anode}} - \Delta G_{\mathrm{cathode}}$$

Recalling Eq. (2.19), where $W_{\mathrm{rev}} = E_{\mathrm{eq}}(n_eF)$, the electrode and full cell equilibrium potential (or open circuit potential) can be obtained respectively as follows:

$$\begin{aligned} E_{\mathrm{eq,hr}} &= -\frac{\Delta G_{\mathrm{hr}}{}^{\Theta}}{(n_eF)} + \frac{RT}{(n_eF)}\ln\left(\frac{\prod (b_i^{\nu_i}\gamma_i^{\nu_i})_R}{\prod (b_i^{\nu_i}\gamma_i^{\nu_i})_P}\right) \\ &= -E^{\circ}_{\mathrm{eq,hr}} + \frac{RT}{(n_eF)}\ln\left(\frac{\prod (b_i^{\nu_i}\gamma_i^{\nu_i})_R}{\prod (b_i^{\nu_i}\gamma_i^{\nu_i})_P}\right) \end{aligned} \tag{2.32}$$

$$\begin{aligned} E_{\mathrm{eq,or}} &= \frac{\Delta G_{\mathrm{anode}}{}^{\Theta}}{(n_eF)} - \frac{\Delta G_{\mathrm{cathode}}{}^{\Theta}}{(n_eF)} + \left(\frac{RT}{(n_eF)}\ln\left(\frac{\prod (b_i^{\nu_i}\gamma_i^{\nu_i})_R}{\prod (b_i^{\nu_i}\gamma_i^{\nu_i})_P}\right)\right)_{\mathrm{cathode}} \\ &\quad - \left(\frac{RT}{(n_eF)}\ln\left(\frac{\prod (b_i^{\nu_i}\gamma_i^{\nu_i})_R}{\prod (b_i^{\nu_i}\gamma_i^{\nu_i})_P}\right)\right)_{\mathrm{anode}} \\ &= E^{\circ}_{\mathrm{eq,anode}} - E^{\circ}_{\mathrm{eq,cathode}} + \left(\frac{RT}{(n_eF)}\ln\left(\frac{\prod (b_i^{\nu_i}\gamma_i^{\nu_i})_R}{\prod (b_i^{\nu_i}\gamma_i^{\nu_i})_P}\right)\right)_{\mathrm{cathode}} \\ &\quad - \left(\frac{RT}{(n_eF)}\ln\left(\frac{\prod (b_i^{\nu_i}\gamma_i^{\nu_i})_R}{\prod (b_i^{\nu_i}\gamma_i^{\nu_i})_P}\right)\right)_{\mathrm{anode}} \end{aligned} \tag{2.33}$$

Here, $E_{\mathrm{eq,hr}}$ is the electrode equilibrium potential (i.e., equilibrium potential for anodic or cathodic half-reaction) and $E_{\mathrm{eq,or}}$ is the full cell equilibrium potential (equilibrium potential for overall reaction including anodic and cathodic half-reactions). The terms $E^{\circ}_{eq,anode}$ and $E^{\circ}_{eq,cathode}$ are the standard potentials for the anode and cathodes, respectively. Eqs. (2.32) and (2.33) are known as *Nernst equations*; they relate the potential of reaction (overall reaction or half reaction) to the standard electrode potential, temperature and activities of species interacting in the reaction. Table 2.1 presents the standard potential for half-reactions occurring in several types of batteries.

2.4 Energy balance in battery

In this section, the general first law of thermodynamics, as expressed in Eqs. (2.8) and (2.9), is applied for a battery system in which the cell temperature is uniform but changes with time. This condition implies that that the battery operates under unsteady state (transient) conditions. The energy balance for a battery can be used to predict the heat generation within the cell and the temperature of the cell, which are important parameters in the thermal management and design of battery systems.

Recall Figs. 2.1 and 2.2, which shows a general battery scheme in which energy is exchanged and generated. In this case, there is no cooling system to hold the battery temperature at a specific value. So, the generated energy within the battery causes the battery cell temperature to rise. This means that the internal energy of the battery increases with time and the cell losses heat to the surroundings (see Fig. 2.3). An energy balance for this battery system, based on Fig. 2.3, can be written as:

$$-\dot{Q}_{loss} - \dot{W} = \frac{dE}{dt} \tag{2.34}$$

where $\dot{Q}_{loss}$ is the rate of heat loss through convection and radiation (the two modes of heat transfer through which the battery loses heat), $\dot{W}$ is the electrical work rate generated during discharge or consumed during charge. The latter rate it is equal to the product of the cell current and cell potential (IV). The term dE/dt is the energy change of the battery cell with respect to time.

The rate of energy loss from the surface of the battery cell involves the convective and radiative heat transfer rates, as follows:

$$\dot{Q}_{loss} = h_0 A(T - T_a) + \varepsilon F A\left(T^4 - T_a^4\right) \tag{2.35}$$

Here, A denotes battery surface area, h_0 heat transfer coefficient, T battery temperature, and T_a ambient temperature. Also, ε is the surface emissivity and F the view factor between the surface of the battery and its surroundings.

Table 2.1 Standard potential for half reactions occurring in several types of batteries.

Battery type	Negative electrode half reaction and its standard potential	Positive electrode half reaction and its standard potential
Lead acid	$Pb(s) + SO_4^{2-}(aq) \rightarrow PbSO_4(s) + 2e^-$ $E^\circ_{anode} = -0.36$ V	$PbO_2(s) + SO_4^{2-}(aq) + 4H^+ + 2e^- \rightarrow PbSO_4(s) + 2H_2O(l)$ $E^\circ_{cathode} = 1.69$ V
Alkaline	$Zn(s) + 2OH^-(aq) \rightarrow ZnO(s) + H_2O(l) + 2e^-$ $E^\circ_{anode} = -1.28$ V	$2MnO_2(s) + H_2O(l) + 2e^- \rightarrow Mn_2O_3(s) + 2OH^-(aq)$ $E^\circ_{cathode} = 0.15$ V
Nickel-zinc	$Zn + 4OH^- \rightarrow ZnOH_4^- + 2e^-$ $E^\circ_{anode} = -1.2$ V	$2NiO(OH) + 2H_2O + 2e^- \rightarrow 2Ni(OH)_2 + 2OH^-$ $E^\circ_{cathode} = 0.5$ V
Lithium-cobalt oxide	$Li^+(aq) + C_6(s) + e^- \rightarrow LiC_6$ $E^\circ_{anode} = -3.0$ V	$CoO_2(s) + Li^+(aq) + e^- \rightarrow LiCoO_2$ $E^\circ_{cathode} = 1.0$ V

Data adapted from Libertexts, 2020. Electrochemistry. Available from: https://chem.libretexts.org/Bookshelves/Analytical_Chemistry/Supplemental_Modules_(Analytical_Chemistry)/Electrochemistry (Accessed 25 May 2022); Bard, A., 2017. Standard Potentials in Aqueous Solution. Routledge, New York.

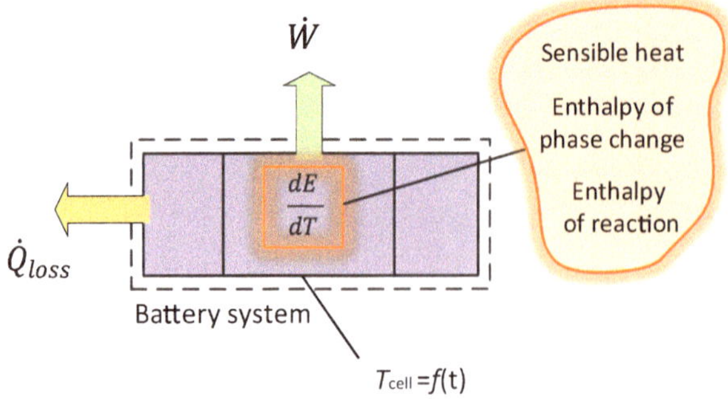

Fig. 2.3 Operation of battery system in the absence of a cooling system (or thermal management system). The temperature of the battery rises with time due to change in the internal energy during the operation of battery.

The rate of change in the energy of battery cell with respect to time (dE/dt) during the electrochemical reactions is due to the change in the state and change in the chemical composition. In other words, sensible energy, latent energy and chemical energy of the battery cell change during the battery operation. Thus, the energy change of the battery with time can be written as:

$$\frac{dE}{dt} = \frac{dE_{\text{senible}}}{dt} + \frac{dE_{\text{latent}}}{dt} + \frac{dE_{\text{chemical}}}{dt} \tag{2.36}$$

Recalling the assumption of uniform temperature throughout the cell and further assuming a negligible change in the heat capacity of the battery during its operation, the rate of change in the sensible energy of battery can be written as:

$$\frac{dE_{\text{senible}}}{dt} = mc_p \frac{dT}{dt} \tag{2.37}$$

where m is the mass of the battery cell and c_p is the specific heat capacity at constant pressure, which is practically constant during battery operation. Also, dT/dt is the change of battery cell temperature with respect to time.

2.4.1 Phase change term

The rate of change of the latent heat of the battery, dE_{latent}/dt, is the average rate of enthalpy change for the phase change process. For instance, in a LiCl-KCl solution, the LiCl species is typically present in the molten electrolyte phase, but there may be partial precipitation of the LiCl due to phase change.

Another example is the crystallization of water as ice in liquid batteries due to low temperature operation. Bernardi et al. (1985) developed the following expression for the average rate of enthalpy change due to phase change:

$$\frac{dE_{\text{latent}}}{dt} = \frac{dH_{\text{phase change}}}{dt} = -\sum_{j\neq m}\sum_{i}\left[\frac{dN_{i,j}}{dt}\left(\Delta h^{\circ}_{i,j\rightarrow m} - RT^2\frac{d\ln\left(\frac{\gamma^{\text{ave}}_{i,m}}{\gamma^{\text{ave}}_{i,j}}\right)}{dt}\right)\right] \tag{2.38}$$

where $\Delta h^{\circ}_{i,\ j\rightarrow m}$ is the molar enthalpy of species i in the secondary reference state corresponding to phase m. The term $dH_{\text{phase change}}/dt$ is calculated based on the average composition of all species. Note that when m phase types are the only phases present during operation of the battery, the enthalpy change due to phase change is zero (i.e., $dH_{\text{phase change}}/dt = 0$).

2.4.2 Enthalpy of reaction term

The change of the chemical energy of the battery system with time, dE_{chemical}/dt, is equivalent the rate of enthalpy change due to the electrochemical reactions in the battery cell. Recalling Eq. (2.5) for the enthalpy of an electrode reaction, the rate of enthalpy change due to the electrochemical reactions can be expressed based on the molar flow rate of reactants ($\dot{N}_P$) and products ($\dot{N}_P$) as:

$$\frac{dH_r}{dt} = \sum_P \dot{N}_P\bar{h}_P - \sum_R \dot{N}_R\bar{h}_R \tag{2.39}$$

Here, dH_r/dt is calculated based on the average composition of all species. The Gibbs-Helmholtz equation can be used to relate enthalpy with Gibbs free energy and temperature as follows:

$$\left(\frac{\partial\left(\frac{\bar{g}}{T}\right)}{\partial T}\right)_P = -\frac{\bar{h}}{T^2} \tag{2.40}$$

Also, the mole flow rate of species i ($\dot{N}_i$) can be expressed based on the Faraday's law which relates the molar transfer rate of a species with current I. The electrode reactions are written so that species i is always in phase m during cell operation and any precipitation of that species in the electrolyte is accounted for separately. So, the mole rate of species i that is produced or consumed in phase m and precipitates in another phase in the electrode reactions can be expressed as

$$\dot{N}_i = \left(\frac{dN_i}{dt}\right)_{j=m} = \left(\frac{\nu_i I}{n_e F}\right)_{j=m} + \left(\frac{dN_i}{dt}\right)_{j\neq m} \tag{2.41}$$

where ν_i is the stoichiometry coefficient of species i, n_e is the number of electrons that are transferred, and F is Faraday's constant ($F = 96{,}487$ C/mole). Subscript j denotes phase. In Eq. (2.41), the second term represents the phase change process that results in precipitation of species i. In the following, this term is accounted for separately as the phase change term. So, the mole rate of species i that is produced or consumed in phase m is $\dot{N}_i = \nu_i I / n_e F$.

Substituting Eqs. (2.40) and (2.41) into Eq. (2.39) and recalling Eq. (2.19) in which the change in the Gibbs free energy of reaction is related to the equilibrium potential (i.e., $-\Delta G = E_{eq}(n_e F)$), the final form for the rate of energy change associated with the enthalpy of a reaction r can be written as

$$\frac{dH_r}{dt} = I_r T^2 \frac{d\frac{E_{eq,r}}{T}}{dT} \tag{2.42}$$

where $E_{eq,r}$ is the equilibrium potential (or theoretical open circuit potential) for electrode reaction r at the average composition. Note that $E_{eq,r}$ can be determined via the Nernst equation presented in Eq. (2.32). The partial current of electrode reaction r, I_r, is negative for an anodic reaction and positive for a cathodic reaction. Note that several electrode reactions take place simultaneously in a battery cell. The total rate of energy change associated with all electrode reactions in the cell is the summation of dH_r/dt associated with each reaction:

$$\frac{dE_{chemical}}{dt} = \frac{dH_{rxn}}{dt} = \sum_r I_r T^2 \frac{d\frac{E_{eq,r}}{T}}{dT} \tag{2.43}$$

Here the term that is multiplied by the partial current is known as enthalpy voltage of reaction r.

2.4.3 Mixing effect

Since $\Delta\dot{H}_{rxn}$ and $\Delta\dot{H}_{phase\ change}$ are defined based on the average composition of all species, there is an error associated with neglecting the variation in composition and mixing effect. To account for these effects, the term of enthalpy of mixing is also added to Eq. (2.36). Substituting Eqs. (2.37), (2.38), and (2.43) into Eq. (2.36) and accounting mixing effect in Eq. (2.36) yields:

$$\frac{dE}{dt} = mc_p \frac{dT}{dt} + \frac{dH_{\text{phase change}}}{dt} + \frac{dH_{\text{rxn}}}{dt} + \frac{dH_{\text{mixing}}}{dt} \tag{2.44}$$

The term dH_{mixing}/dt denotes the rate of enthalpy change due to the mixing process is represented by the variation of enthalpy with composition. The enthalpy of mixing includes the relaxation or generation concentration profile (Bernardi et al., 1985). In the absence of phase change effects, this term denotes the heat generation rate due to the interruptions in the current during operation of the battery. The relaxation effect occurs after the cessation of discharge current in a battery. When the discharge current is stopped the voltage jumps at first and then gradually rises over many hours. The heat is generated during relaxation due to the concentration gradients that are formed during the passage of current (i.e., generation of concentration profile). This heat generation is associated with the heat of mixing. During relaxation of the battery, the concentration gradients gradually diminish over time, tending to a uniform composition concentration profile. Bernardi et al. (1985) derived the following expression for dH_{mixing}/dt:

$$\frac{dH_{\text{mixing}}}{dt} = -\sum_j \frac{d}{dt}\left[\int_{v_j} \sum_i c_{i,j} RT^2 \frac{\partial \ln\left(\frac{\gamma_{i,j}}{\gamma_{i,j}^{\text{ave}}}\right)}{\partial T} dv_j\right] \tag{2.45}$$

where dv_j is the differential volume element of phase j and $c_{i,\,j}$ is the concentration of species i in phase j. It can be seen from Eq. (2.45) that the rate of enthalpy change of mixing is a function of the spatial change in the composition, while other terms like $dH_{\text{phase change}}/dt$ and dH_{rxn}/dt are dependent on the average composition. The analytical calculation of dH_{mixing}/dt is difficult as it requires complicated integrations over concentration profiles. However, simplifying by neglecting the enthalpy of mixing can result in underestimation or overestimation of battery temperature. For example, the temperature of a lead-acid battery that is properly insulated usually increases after cessation of discharge current, while for a high-temperature battery the temperature decreases after the interruption of current. Also, during battery operation (passage of current) where the concentration profile is generated, the temperature is usually underestimated if the mixing effect is not considered. For the conditions in which the enthalpy of mixing can be neglected, the average composition should be considered wisely to minimize the errors associated with neglecting of the mixing effect.

2.5 Heat generation rate in batteries

In the case when mixing and phase change effects can be neglected in the operation of a battery cell and convective heat transfer is the only contributor to the heat loss rate, the balance in Eq. (2.34) can be expressed as follows:

$$mC_p \frac{dT}{dt} + \sum_{r=1,2,\ldots} I_r T^2 \frac{d\frac{E_{\mathrm{eq},r}}{T}}{dT} = -h_0 A(T - T_a) - IV \tag{2.46}$$

The term for rate of enthalpy of reaction can be written as the summation $-\sum_{r=1,2,\ldots} I_r E_{\mathrm{eq},r}$ and $\sum_{r=1,2,\ldots} I_r T dE_{\mathrm{eq},r}/dT$. Thus, Eq. (2.46) can be re-written as:

$$h_0 A(T - T_a) + mc_p \frac{dT}{dt} = \left(\sum_{r=1,\ 2,\ \ldots} I_r E_{\mathrm{eq},r} - IV \right) - \sum_{r=1,2,\ldots} I_r T \frac{dE_{\mathrm{eq},r}}{dT} \tag{2.47}$$

Here, the right side represents the *heat generation rate* within a battery system, which is the summation of the rate of change of heat storage in the system ($mc_p dT/dt$) and the heat transfer rate out of the system ($h_0 A(T - T_a)$). Recalling that the reversible work rate is IE_{eq}, the term in parentheses on the right side is seen to be the difference between the real and reversible potentials. This gives the cell overpotential due to irreversibilities in the electrochemical reactions. This term is known as the polarization heat rate (or irreversible heat rate) and it includes the ohmic heat (or joule heat) and charge and mass transfer overpotentials. When the battery cell operates reversibly, this term is zero (i.e., $\sum_{r=1,2,\ldots} I_r E_{\mathrm{eq},r} - IV = 0$). The ohmic heat is caused by current transfer across an internal resistance.

The term $\sum_{r=1,2} I_r T\left(dE_{\mathrm{eq},r}/dT\right)$, which is embodied in the enthalpy of reaction, is known as the entropic or reversible heat rate. This term indicates that heat is generated due to reversible entropy change in the cell. In fact, this term denotes the heat generation rate over reversible isothermal operation of the battery cell. The term is called entropic heat as it is proportional to the entropy change associated with the electrochemical reactions, i.e.,

$$\sum_{r=1,2,\ldots} I_r T\left(dE_{\mathrm{eq},r}/dT\right) = T \sum_{r=1,2,\ldots} I_r dE_{\mathrm{eq},r}/dT = T \sum_{r=1,2,\ldots} \frac{1}{n_{e,r}F} dS_r \tag{2.48}$$

Therefore, the source of heat generation in a battery cell is comprised of reversible and irreversible processes that are determined by the reversible heat (entropic heat) and irreversible heat (polarization heat) caused by ohmic loss, charge transfer and mass transfer irreversibilities. For the condition where mixing and phase change effects are also taken into account, as in Eq. (2.47), the rate of energy generation within a battery can be written as:

$$\dot{Q}_{gen} = \left(\sum_{r=1,\,2,\,\ldots} I_r E_{eq,r} - IV \right) - \sum_{r=1,2,\ldots} I_r T \frac{dE_{eq,r}}{dT} + \frac{dH_{mixing}}{dt} + \frac{dH_{phase\ change}}{dt} \tag{2.49}$$

During discharge of a battery, the active materials in electrodes are converted to discharge products through electrochemical reactions until the voltage of the battery approaches the voltage threshold. The extent of this conversion is known as the *utilization percentage of electrode material*. Also, the utilization percentage indicates the ratio of charge that is used after passing a specific time during discharge to the initial amount of charge (i.e., at the beginning of the discharge process).

Table 2.2 shows the magnitude of the terms in the energy balance equation (Eq. 2.49) at several utilization percentage of active material levels for the electrode in a rechargeable lithium-aluminum/iron sulfide (LiAl-FeS) battery, based on a commonly cited source. The following electrochemical reactions occur simultaneously in the positive electrode of this battery:

Table 2.2 Heat rate contributions (in W/m^2) including reversible and irreversible heat generation rates for positive electrode of LiAl-FeS battery.

	Utilization percentage (%)			
Energy rate contribution	**0**	**10**	**20**	**30**
Polarization heat rate	33	45	50	55
Entropic heat rate	7	7	7	7
Enthalpy of reaction and electrical work rate (i.e., summation of first and second rows)	40	52	57	62
Mixing heat rate	0	0.05	0.5	0
Heat loss rate	−37	−37	−37	−37

Data from Bernardi, D., Pawlikowski, E., Newman, J., 1985. A general energy balance for battery systems. J. Electrochem. Soc. 132(1), 5–12.

$$\text{Reaction 1 } (r=1): 2FeS + 2Li + 2e^{-} \rightarrow Li_2FeS_2 + Fe \quad (2.50)$$

$$\text{Reaction 2 } (r=2): Li_2FeS_2 + 2Li^{+} + 2e^{-} \rightarrow 2Li_2S^{-} + Fe \quad (2.51)$$

The battery cell is insulated properly so that the cell temperature and heat loss rate do not change significantly (i.e., isothermal operation). It can be seen that the enthalpy of mixing effect is much smaller than the enthalpy of reaction. In addition, the polarization heat rate increases at higher utilization percentages, while the entropic heat rate remains constant.

For isothermal operation of a battery, all generated heat within the battery is transferred outside of it (e.g., to a thermal management system). An isothermal model of a battery provides useful information to understand its operation at various temperatures and how temperature influences battery performance. With such a model, a temperature range can be identified over which the battery operates efficiently. To maintain the battery temperature in the operational temperature range, thermal management systems are designed to control the temperature of the battery temporally and spatially. Battery thermal management systems are described in detail in Chapter 5.

In recent years many areas related to battery systems and battery thermal management systems have been the subject of investigations, assessments and advances. These include techniques for novel thermal management systems for batteries (Al-Zareer et al., 2018, 2019a; Malik et al., 2016; Koohi-Fayegh and Rosen, 2020), and electrochemical modeling and thermal performance management systems for lithium-ion battery (Al-Zareer et al., 2017, 2019b, 2020; Malik et al., 2018). Further studies are still required to develop battery system applications and thermal management systems for batteries in diverse sectors, especially transportation, to enhance energy utilization and reduce environmental impacts in terms of carbon and pollutant emissions.

2.6 Closing remarks

This chapter describes the development of a thermodynamic model for a battery cell. First, thermodynamic fundamentals are described to help in understanding battery thermodynamics. The first and second laws of thermodynamics are applied to a generalized battery cell. From these, it is seen that the cell potential is related to the Gibbs free energy. The reversible operation of a battery is assumed to derive the equilibrium potential and Nernst equation. Then, the real (irreversible) operation of a battery is considered, and the heat generation rate within the battery and overpotentials due to irreversibilities in electrochemical reactions are identified. With the thermodynamic model of a battery, a temperature range for efficient operation of the battery can be determined. This makes it clear that the design of battery thermal management systems to maintain battery temperature in the optimum range is important.

Study questions

2.1. Define the terms enthalpy of reaction and Gibbs free energy of reaction.

2.2. What is reversible work? How is it related to the reversible electrical potential of a battery?

2.3. What is the difference between work and reversible work for a battery?

2.4. How is the reversible potential of reaction related to the standard electrode potential, temperature and activities of species interacting in the reaction?

2.5. What is the general equation for the heat generation rate in a battery cell?

2.6. What is reversible heat rate in a battery system? How does it differ from irreversible heat rate in a battery system?

2.7. Consider a battery cell that is insulated properly so that the cell temperature is kept constant at 30°C. The battery can generate 100 kW of electrical power during discharge, in which 130 kJ/s of heat is generated in the battery. For the case when mixing and phase change effects can be neglected in the operation of the battery cell and convective heat transfer is the only contributor to the heat loss rate, determine the heat loss rate from the battery?

2.8. Consider a lithium cobalt oxide battery cell with a mass of 100 g and surface area of 100 cm^2 in which the redox reactions at the anode and cathode during discharging can be written as follows:

$$\text{Oxidation at anode}: LiC_6 \rightarrow C_6 + e^- + Li^+$$

$$\text{Reduction at cathode}: CoO_2 + e^- + Li^+ \rightarrow LiCoO_2$$

$$\text{Overall reaction}: LiC_6 + CoO_2 \rightarrow C_6 + LiCoO_2$$

In the operation of the lithium cobalt battery, mixing and phase change effects can be neglected. Also, the lithium cobalt battery is exposed to air (as the coolant) at a temperature of 20°C. The equilibrium potential can be expressed as a function of temperature, as follows:

$$E_{eq} = -\frac{1}{84}T^2 + \frac{1}{24}T - 5.25 \quad \left(\text{for } E_{eq} \text{ in V and } T \text{ in K}\right)$$

Assuming steady state operating conditions for the battery system, as well as a natural convection heat transfer coefficient of 20 W/m^2 K between the surface of the battery and the air, determine the voltage of the battery during discharge as a function battery temperature when the current of the battery is 12 A.

2.9. Illustrate the lithium ion battery voltage variations with battery temperature in Study Question 2.8 when the temperature of the battery varies between 20°C and 30°C.

References

Al-Zareer, M., Dincer, I., Rosen, M.A., 2017. Electrochemical modeling and performance evaluation of a new ammonia-based battery thermal management system for electric and hybrid electric vehicles. Electrochim. Acta 247, 171–182.

Al-Zareer, M., Dincer, I., Rosen, M.A., 2018. A review of novel thermal management systems for batteries. Int. J. Energy Res. 42 (10), 3182–3205.

Al-Zareer, M., Dincer, I., Rosen, M.A., 2019a. Comparative assessment of new liquid-to-vapor type battery cooling systems energy. Energy 188, 116010.

Al-Zareer, M., Dincer, I., Rosen, M.A., 2019b. A novel approach for performance improvement of liquid to vapor based battery cooling systems. Energy Convers. Manage. 187, 191–204.

Al-Zareer, M., Dincer, I., Rosen, M.A., 2020. A thermal performance management system for lithium-ion battery packs. Appl. Therm. Eng. 165, 114378.

Bernardi, D., Pawlikowski, E., Newman, J., 1985. A general energy balance for battery systems. J. Electrochem. Soc. 132 (1), 5–12.

Dincer, I., Rosen, M.A., 2021. Thermal Energy Storage: Systems and Applications, third ed. Wiley, London.

Farsi, A., Zamfirescu, C., Dincer, I., Naterer, G.F., 2020. Electrochemical transport in CuCl/HCl (aq) electrolyzer cells and stack of the Cu–Cl cycle. J. Electrochem. Soc. 167 (4), 044515.

Johnson, J.W., Oelkers, E.H., Helgeson, H.C., 1992. SUPCRT92: a software package for calculating the standard molal thermodynamic properties of minerals, gases, aqueous species, and reactions from 1 to 5000 bar and 0 to 1000°C. Comput. Geosci. 18 (7), 899–947.

Koohi-Fayegh, S., Rosen, M.A., 2020. A review of energy storage types, applications and recent developments. J. Energy Storage 27, 101047.

Malik, M., Dincer, I., Rosen, M.A., 2016. Review on use of phase change materials in battery thermal management for electric and hybrid electric vehicles. Int. J. Energy Res. 40 (8), 1011–1031.

Malik, M., Dincer, I., Rosen, M.A., Mathew, M., Fowler, M., 2018. Thermal and electrical performance evaluations of series connected Li-ion batteries in a pack with liquid cooling. Appl. Therm. Eng. 129, 472–481.

Newman, J., Thomas-Alyea, K.E., 2012. Electrochemical Systems, third ed. John Wiley & Sons, New Jersey.
Rosen, M., Farsi, A., 2022. Sustainable Energy Technologies for Seawater Desalination. Elsevier, Academic Press, United Kingdom.
Zamfirescu, C., Naterer, G.F., Rosen, M.A., 2017. Chemical exergy of electrochemical cell anolytes of cupric/cuprous chlorides. Int. J. Hydrog. Energy 42 (16), 10911–10924.

CHAPTER 3

Electrochemical modeling of batteries

Chapter objectives

- Develop an electrochemical model for a generalized battery system.
- Discuss the irreversibilities caused by the overpotentials in a battery system.
- Determine the heat generation rate in a battery system due to the electrochemical reactions, and ion and charge transfer processes.
- Discuss the performance of a battery by defining the energy efficiency for a battery system and other performance measures.

Nomenclature

Symbols

B	exponential factor
c	concentration (mol/cm^3)
D	diffusion coefficient (cm^2/s)
E	cell potential (V)
F	Faraday constant (coulomb/mol)
g	specific Gibbs free energy (J/mol)
Δg_a	specific Gibbs function of activation (J/mol)
H	height of battery (cm)
i	current density (A/cm^2)
i_0	exchange current density (A/cm^2)
I	total current (A)
j	molar flux density (mol/cm^2 s)
k_c	reaction rate constant
n_e	number of transferred electrons in electrode reaction
r	radial position (cm)
R	universal gas constant (8.3143 J/mole K)
T	temperature (°C, K)
t	time (s)
tr	transference number
u_i	mobility of species i (cm^2 mol/J s)
v	velocity of fluid (cm/s)
z	charge number

Battery Technology
https://doi.org/10.1016/B978-0-443-18862-6.00007-0

Greek letters

α	transfer coefficient of reaction
β	symmetry factor
η_{energy}	energy efficiency
η_c	concentration overpotential (V)
η_{ov}	total overpotential (V)
η_{ohm}	ohmic overpotential (V)
η_s	surface overpotential (V)
Φ	potential (V)
κ	ionic conductivity (Siemens per meter (S/m))
γ	activity coefficient
σ	electronic conductivity (S/cm)

Subscripts

a	anode
c	cathode
conc	concentration
eq	equilibrium
j	phase indicator
ocp	open circuit potential

3.1 Introduction

The electrochemical modeling of batteries is described in this chapter. Included is an electrochemical model of battery systems. The battery electrochemical model encompasses electrode electrochemical reactions, ion transport through the electrolyte, diffusion and charge transfer.

Transport, thermodynamics, and kinetics are interrelated in electrochemical studies. With an electrochemical model of a battery cell, the overpotentials within the cell can be determined, which subsequently can be used to calculate heat generation in a cell. In other words, the heat generation rates in electrochemical reactions during charging/discharging and Joule heating (heat generation due to internal resistance to the current flow) can be determined with the electrochemical model. Then these heat generation rates are used in the energy balance (presented in Chapter 2) to determine the temperature of battery. The temperature of battery can be also determined through the heat and mass transfer model, the results of which are transferred to the electrochemical model to update the electrochemical calculations.

Thus, the heat generation rate within a battery and the temperature of the battery are two parameters that connect the electrochemical model with the heat and mass transfer model. In the following section, the overall cell potential is presented. Then, the electrochemical reactions, charge and ion transport phenomena through an electrolyte, and the resulting overpotentials in a battery cell are described.

3.2 Overall cell potential of a battery

The electrochemical analysis allows us to determine the potential losses in an electrochemical cell-like battery (Farsi et al., 2020). Identifying various types of overpotentials and losses helps in determining the reasons and locations for their occurrences and ways to overcome these irreversibilities for a given battery. In a battery system, the overpotentials mainly include surface overpotential, ohmic overpotential, and concentration overpotential. The surface overpotential is due to the resistance to electrochemical reactions, while the ohmic overpotential is due to the resistance to electronic and ionic current, and the concentration overpotential is due to the presence of a concentration gradient in an electrolyte.

It is noted that the equilibrium potential (or reversible potential) is the maximum cell potential that is attained in reversible operation of a battery. The actual cell potential of the battery is lower than the equilibrium potential due to the overpotentials that are induced by the passage of current. The potential of a battery cell is also a maximum when no current passes across the cell. This potential is referred to as the open circuit potential, and it is almost equal to the equilibrium potential.

The overpotentials in a battery cell are the irreversibilities in real operation of the battery. So, by subtracting the equilibrium potential (E_{eq}) from the real potential (E), the overpotentials (η_{ov}) can be evaluated as:

$$\eta_{ov} = E - E_{eq} \tag{3.1}$$

where η_{ov} is the total overpotential of a cell. To split the total overpotential into its components, including surface overpotential, ohmic overpotential and concentration overpotential, consider Fig. 3.1 which shows a generalized battery cell. The cell is comprised of an anode and a cathode (working electrodes). Each of these has two reference electrodes which are placed near the working electrodes (points 1 and 2) and at the bulk region where there is a uniform concentration (point 3). The reference electrodes are made of the same materials as the corresponding working electrodes. For instance, if the

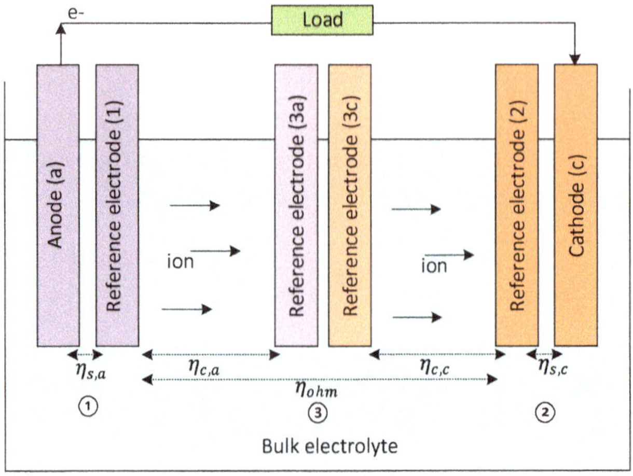

Fig. 3.1 Working electrodes, reference electrodes and the region of overpotentials in a battery cell.

anode is made from graphitic carbon, its corresponding reference electrodes (placed near and at bulk region) are also made up from graphitic carbon. No current flows from the reference electrodes and they only sense the potential in the electrolyte that is carrying current.

To calculate the potential difference between the anode and cathode (i.e., real cell potential, E), we can move from the anode toward the cathode accounting the potential differences between each of the working/reference electrodes that are in between. Then, the potential difference between the anode and cathode can be generally written as:

$$E = \Phi_a - \Phi_c = (\Phi_a - \Phi_1) + (\Phi_1 - \Phi_2) - (\Phi_c - \Phi_2) \quad (3.2)$$

Here, $(\Phi_1 - \Phi_2)$ can be further divided into:

$$(\Phi_1 - \Phi_2) = (\Phi_1 - \Phi_{3,a}) + (\Phi_{3,a} - \Phi_{3,c}) + (\Phi_{3,c} - \Phi_2) + \eta_{ohm} \quad (3.3)$$

where Φ_a and Φ_a are the anode and cathode potentials respectively, Φ_1 and $\Phi_{3,a}$ are the potentials associated with the reference electrodes for the anode that are placed close to anode (point 1) and at the bulk (point 3), respectively; Φ_2 and $\Phi_{3,c}$ are the potentials associated with the reference electrodes for cathode that are placed close to cathode (point 2) and at bulk (point 3), respectively; and η_{ohm} is the ohmic overpotential due to the passage of current through the electrolyte.

As stated before, the surface resistance is associated with the resistance to electrochemical reactions. In fact, a driving force (activation energy) is required to induce electrochemical reactions to occur. This driving force is referred to as surface overpotential. To measure this surface overpotential, a reference electrode needs to be placed near its working electrode. Since at the reference electrode there is no current and at the working electrode the potential induces the current to flow, the potential difference between this working electrode and its reference electrode measures the surface overpotential. So, the first and third terms on the right sides of Eq. (3.2) represent the surface overpotentials at the anode ($\eta_{s,a}$) and the cathode ($\eta_{s,c}$), respectively, i.e.,

$$(\Phi_a - \Phi_1) = \eta_{s,a} \tag{3.4}$$

$$(\Phi_c - \Phi_2) = \eta_{s,c} \tag{3.5}$$

Surface overpotential is discussed in detail in the next section.

As noted earlier, the concentration overpotential is due to the concentration gradient in an electrolyte. The concentration variations within an electrolyte need to be measured after stopping the current and before the beginning of concentration distribution changes due to diffusion and convection. The reference electrodes placed in the bulk region assess the conditions where the concentration is distributed uniformly and no current is flowing. So, the potential difference between these reference electrodes and the other reference electrodes placed near the working electrodes represents the overpotential due to concentration difference. Thus, in Eq. (3.3), the first and third terms correspond to the concentration overpotentials at the anode ($\eta_{c,a}$) and the cathode ($\eta_{c,c}$), respectively, i.e.,

$$(\Phi_1 - \Phi_{3,a}) = \eta_{c,a} \tag{3.6}$$

$$(\Phi_{3,c} - \Phi_2) = \eta_{c,c} \tag{3.7}$$

The concentration overpotential is defined through the transport phenomena in a cell, which is discussed in Section 3.4.

The equilibrium potential is defined as the potential at equilibrium conditions where there is no current and no concentration gradient to cause irreversibilities. Since the two reference electrodes are placed in the bulk fluid where uniform concentrations are present, the difference between these two reference electrodes is equal to the reversible potential. Consequently, the second term on the right side of Eq. (3.3) is equivalent to the cell equilibrium potential (or open circuit potential), i.e.,

$$(\Phi_{3,a} - \Phi_{3,c}) = E_{eq} \tag{3.8}$$

In the following sections, the irreversibilities in a battery system that are associated with electrochemical reactions and ion and charge transfer processes are discussed in detail.

3.3 Surface overpotential

As noted earlier, the surface overpotential is the resistance to electrochemical reactions. In other words, this irreversibility is related to the activation energy of the rate-determining step in the electrochemical reaction. The rate-determining step is the slowest step of the electrochemical reaction and it usually controls the rate of the electrochemical reaction. On the surface of an electrode, several random thermal collisions cause oxidation and reduction reactions to take place in forward and backward directions. However, it is assumed for simplicity that one elementary step entails the slowest electrochemical reaction and thus controls the reaction rate in a battery cell, as follows:

$$O + n_e e^- \rightleftharpoons R \tag{3.9}$$

where *O* denotes the oxidized species and *R* the reduced species. The forward reaction corresponds to the formation of products and the backward reaction corresponds to the formation of reactants. The reaction at the anode is oxidation and at the cathode is reduction. Fig. 3.2 shows the relative energy of an electron in the oxidation and reduction reactions. During oxidation, the electron is transferred from a relatively high energy of reactant to

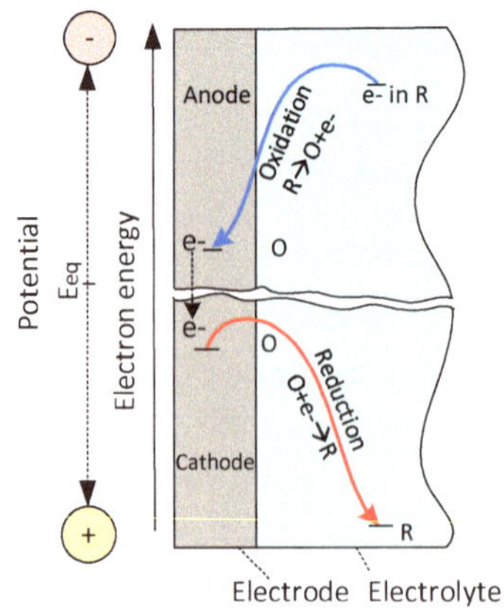

Fig. 3.2 Relative energy of an electron in oxidation and reduction reactions at the electrode/electrolyte interface.

the electrode. During reduction, the electron is transferred from the electrode to the relatively low energy of the reactant. In other words, applying a negative potential in the cathode increases the energy of the electron which causes the reduction of a species near the electron. Also, applying a positive potential to the anode reduces the energy of the electron, which results in transfer of the electron from the reactant to the electrode.

The rate of a half-reaction can be expressed based on the rate of production of R species in the forward reaction or the rate of consumption of R in the backward direction. The rate of reaction can be defined in terms of the current density at the electrode. Thus, the difference between the rate of forward and backward reactions corresponds to the net current density resulting from the anodic and cathodic reactions.

Fig. 3.3 shows a typical control volume near the cathode. Applying a mass balance to a unit area of the interface between the electrode and the electrolyte for the forward reaction indicates that the transport rate of O species from the bulk electrolyte to the electrode is equal to the rate of consumption of O at the forward reaction. Thus, the current density resulting from the cathodic reaction (i_c) can be written as

$$\frac{i_c}{n_e F} = K_c c_O = T B_c \exp\left(\frac{-\Delta g_c}{RT}\right) c_O \tag{3.10}$$

where B_c is the exponential factor for the cathode reaction, T is absolute reaction temperature, R is the universal gas constant, Δg_c is change in the specific Gibbs function during the cathodic reaction, and c_O is the reactant (O) concentration for the cathodic reaction. It is pointed out that Eq. (3.10) is analogous to the expression for a first order chemical reaction.

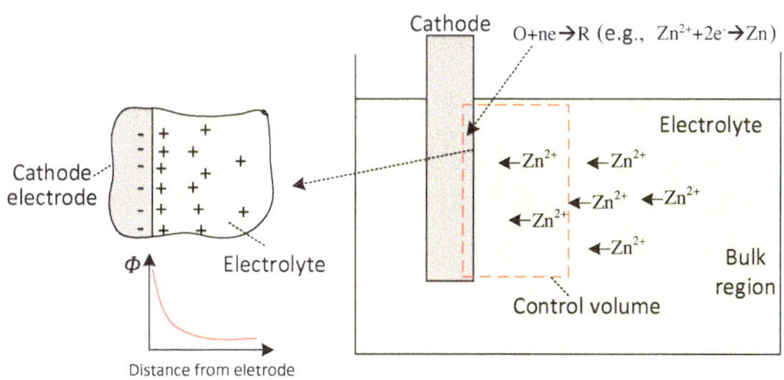

Fig. 3.3 Potential profile and control volume adjacent to a cathode placed in an electrolyte.

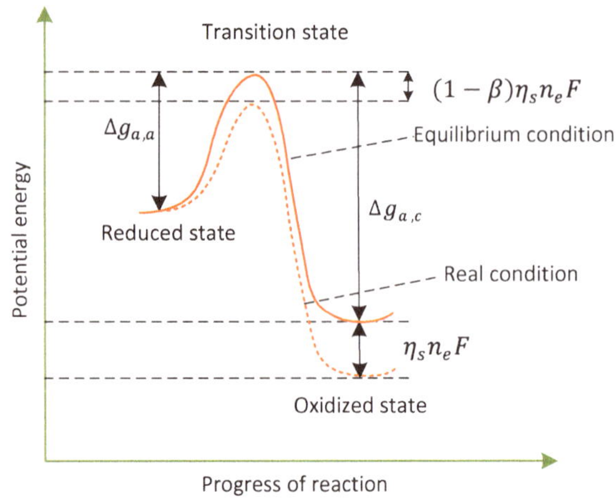

Fig. 3.4 Plot of energy potentials for an elementary charge transfer process ($O+n_e e^- \rightleftharpoons R$). The solid line is associated with the applied potential in real conditions ($\Delta\Phi$) and the dashed line corresponds to the equilibrium potential ($\Delta\Phi_{eq}$).

For both the forward and backward reactions, an energy barrier exists that needs to be crossed for the reaction to begin and proceed. This can be described with transition state theory and is thereby illustrated by a graph plotting potential energy against the degree of completion for a reaction (see Fig. 3.4). For the forward reaction, the reactants need to pass the peak of the curve, which corresponds to the energy barrier. This energy barrier (or activation energy) is equal to the difference between the Gibbs function of the reactant and the activated complex placed at the peak of curve. Similarly, for the backward reaction, the difference between the Gibbs function of the products and the activated complex corresponds to the energy barrier that needs to be overcome.

The change in the specific Gibbs function during cathodic reaction can be expressed as:

$$\Delta g_c = \Delta g_{a,c} + \beta n_e F \Delta \Phi \qquad (3.11)$$

where $\Delta g_{a,c}$ is the specific Gibbs function of activation for the cathode reaction, n_e is number of transferred electrons and F is Faraday's constant. The term $\Delta\Phi$ is the potential difference between the electrode and the electrolyte; this can be approximated as the potential relative to a given electrode (e.g., reference electrode). In fact, the potential decreases moving from the electrode-electrolyte interface to the bulk electrolyte. When an electrode is placed in an electrolyte, two parallel layers of charge are created at the electrode and electrolyte side interface. This parallel layer of charge is referred to as the electric double layer (see Fig. 3.3). Note that, with some simplifications, the electrode/electrolyte interface can be considered as a capacitor.

The parameter β in Eq. (3.11) is the symmetry factor, which typically varies between zero and unity; however, it usually has the value of 0.5. The transfer coefficients for the anodic and cathodic reactions (α_a and α_c, respectively) are other kinetic parameters that can be written in terms of symmetry factor β:

$$\alpha_a = 1 - \beta \tag{3.12}$$

$$\alpha_c = \beta \tag{3.13}$$

Conceptually, the symmetry factor indicates how the electric filed formed by the potential difference favors one direction of reaction over the other. The term $\beta n_e F \Delta\Phi$ in Eq. (3.11) indicates the work done by the O species to overcome the electric field. The electric field for this reaction ($O + n_e e^- \rightarrow R$) opposes the motion of O toward the cathode. Thus, substituting Eq. (3.11) into Eq. (3.10) gives:

$$\frac{i_c}{n_e F} = TB_c \exp\left(-\frac{\Delta g_{a,c}}{RT}\right) \exp\left(-\frac{\beta n_e F \Delta\Phi}{RT}\right) c_O = k_c \exp\left(-\frac{\beta n_e F \Delta\Phi}{RT}\right) \tag{3.14}$$

Here, k_c denotes the rate constant for the cathodic reaction, which is approximated well by the Arrhenius equation (i.e., $k_c = TB_c \exp(-\Delta g_{a,c}/RT)$).

Similarly, for the backward reaction where oxidation occurs at the anode, the current density resulting from the anodic reaction is equal to the rate of consumption of R species. Thus the current density resulting from the anodic reaction (i_a) can be expressed as:

$$\frac{i_a}{n_e F} = K_a c_R = TB_a \exp\left(\frac{-\Delta g_a}{RT}\right) c_R \tag{3.15}$$

Here, Δg_a is change in the specific Gibbs function during the anodic reaction, B_a is the exponential factor for the anodic reaction, and c_R is the reactant (R) concentration for anodic reaction. The term Δg_a can be expressed as:

$$\Delta g_a = \Delta g_{a,a} - (1 - \beta) n_e F \Delta\Phi \tag{3.16}$$

where $\Delta g_{a,a}$ is the specific Gibbs function of activation for the anode reaction. The negative sign before the second term on the right side indicates that the electric field helps the motion of R species from the anode to the bulk electrolyte, thereby decreasing the value of Δg_a. Substituting Eq. (3.15) into Eq. (3.16) yields

$$\frac{i_a}{n_e F} = TB_a \exp\left(-\frac{\Delta g_{a,a}}{RT}\right) \exp\left(-\frac{(1-\beta)n_e F\Delta\Phi}{RT}\right) c_O$$
$$= k_a \exp\left(-\frac{(1-\beta)n_e F\Delta\Phi}{RT}\right) \tag{3.17}$$

Here, k_a denotes the rate constant for the anodic reaction, which is approximated well by the Arrhenius equation (i.e., $k_a = TB_a \exp(-\Delta g_{a,a}/RT)$).

For reversible reaction conditions, the same number of ions and electrons cross the electrolyte and electrode, meaning that the same number of reactants is converted to products and vice versa. This also means that that the current densities associated with the anodic and cathodic reactions are the same [i.e., the net current density is zero (the open circuit condition)]. The potential difference between the electrode and electrolyte interface at the reversible condition is the equilibrium (or reversible) potential $\Delta\Phi_{eq}$. The current density corresponding to this condition is known as exchange current density i_0 and can be written as:

$$i_0 = i_a = i_c \tag{3.18}$$

Using Eqs. (3.14), (3.17), (3.18), the equilibrium potential can be obtained as follows:

$$k_a\left(\exp\left(\frac{(1-\beta)}{RT} n_e F\Delta\Phi_{eq}\right)\right) c_R = k_c\left(\exp\left(\frac{(-\beta)}{RT} n_e F\Delta\Phi_{eq}\right)\right) c_O \tag{3.19}$$

Rearranging the above equation gives

$$\Delta\Phi_{eq} = E_{eq} = \frac{RT}{n_e F} \ln\left(\frac{k_c c_O}{k_a c_R}\right) \tag{3.20}$$

Substituting Eq. (3.20) into Eqs. (3.14), (3.17), (3.18) yields the exchange current density, as follows:

$$i_0 = nFk_a^{\beta} k_c^{1-\beta} c_R^{\beta} c_o^{1-\beta} \tag{3.21}$$

The exchange current density i_0 depends on the temperature, reactant and product concentrations and nature of electrode-electrolyte interface. Generally, the exchange current density varies between 10^{-7} mA/cm^2 and 1 mA/cm^2.

In real operation of a cell where overpotentials exist, the rates for the anodic and cathodic reactions are not the same, resulting in a net flow of current ($i \neq 0$). Thus the net current density resulting from the anodic and cathodic reactions can be written as

$$\begin{aligned}\frac{i}{n_e F} &= \frac{1}{n_e F}(i_a - i_c)\\ &= \left(k_a\left(\exp\left(\frac{(1-\beta)}{RT}n_e F\Delta\Phi\right)\right)c_R\right)\\ &\quad -\left(k_c\left(\exp\left(\frac{(-\beta)}{RT}n_e F\Delta\Phi\right)\right)c_O\right)\end{aligned} \tag{3.22}$$

In the nonequilibrium case, the overpotential increases the oxidation rate and reduces the reduction rate by reducing the oxidized state energy level (see Fig. 3.4). In fact, the activation energy in the cathode direction increases. The potential difference between the electrode and the electrolyte can be written in terms of equilibrium potential difference $\Delta\Phi_{eq}$ and the overpotential η_s, as

$$\Delta\Phi = \Delta\Phi_{eq} + \eta_s \tag{3.23}$$

To understand what happens in the real operating conditions of a cell, consider the anode in which the oxidation reaction of $Zn \rightarrow Zn^{2+} + 2e^-$ occurs, and current is passed through the cell. In these circumstances, the potential drop in the vicinity of the electrode changes from $\Delta\Phi_{eq}$ to a more positive value ($\Delta\Phi$) due to irreversibilities. It becomes more positive with an increase in the current density. Likewise, in the cathode the continuous reduction reaction of $Zn^{2+} + 2e \rightarrow Zn$ with the passage of current causes the potential drop near the electrode to change from $\Delta\Phi_{eq}$ to a more negative value ($\Delta\Phi$) owning to irreversibilities. It also becomes more negative with an increase in the current density. So, when current passes through the cell, the absolute value of the actual potential drop is always larger than the absolute equilibrium potential due to the overpotentials (i.e., $|\Delta\Phi| > |\Delta\Phi_{eq}|$). Moreover, it can be concluded that the surface overpotential is positive for the anodic reaction and negative for the cathodic reaction.

In Fig. 3.4, we show how applying a nonreversible potential $\Delta\Phi$ changes the oxidized state energy level from $\Delta g_{a,a}$ to Δg_a for the anodic direction and from $\Delta g_{a,c}$ to Δg_c for the cathodic direction. In fact, the activation energy for the cathodic direction increases by $\beta\eta_s n_e F$ and for the cathodic direction decreases by $(1-\beta)\eta_s n_e F$. This can be better understood through noting that the overpotential aids the anodic reaction to proceed and deters the cathodic reaction.

We can write Eq. (3.22) in terms of the equilibrium potential and surface overpotential (i.e., $\Delta\Phi = \Delta\Phi_{eq} + \eta_s$). Doing so gives

$$\frac{i}{n_e F} = k_a\left(\exp\left(\frac{1-\beta}{RT}n_e F\left(\Delta\Phi_{eq} + \eta_s\right)\right)\right)c_R - k_c\left(\exp\left(\frac{-\beta}{RT}n_e F\left(\Delta\Phi_{eq} + \eta_s\right)\right)\right)c_O \tag{3.24}$$

Substituting the equilibrium potential expression (Eq. 3.20) into the above equation yields

$$\frac{i}{n_e F} = k_a\left[\exp\left(\frac{1-\beta}{RT}n_e F\eta_s + \alpha_a \ln\left(\frac{k_c c_O}{k_a c_R}\right)\right)\right]c_R - k_c\left[\exp\left(\frac{-\beta}{RT}n_e F\eta_s - \alpha_c \ln\left(\frac{k_c c_O}{k_a c_R}\right)\right)\right]c_O \tag{3.25}$$

Modifying Eq. (3.25) in terms of the exchange current density expression (Eq. 3.21) yields the following expression for the net current density:

$$i = i_0\left(\exp\left(\frac{1-\beta\eta_s n_e F}{RT}\right) - \exp\left(\frac{-\beta\eta_s n_e F}{RT}\right)\right) \tag{3.26}$$

This equation provides the relation between the current density and overpotential and is referred to as the *Butler-Volmer equation*. The exchange current density i_0 in this equation is similar to the reaction rate constant in chemical kinetics. One way to determine i_0 is through a graph of $\ln(i)$ against η_s which can be obtained based on experimental data for specific types of reactants, products, electrodes and electrolytes.

3.4 Concentration overpotential

During the charge or discharge operation of a battery, transfer of mass, momentum and charge and also transfer of heat and work occur simultaneously. The rate of each of these mechanisms affects the performance of battery significantly. Mathematical formulations that are used to describe these transfer processes become complex when considering all these phenomena at the same time. In this section, the transfer processes within a battery cell and the resulting concentration polarization are described briefly.

In the previous section, for deriving an expression for the surface overpotential, it was assumed that the concentrations of species do not change with current density. However, in real operation of a battery, there are concentration gradients in different parts of the electrolyte. These concentration gradients often result from a low concentration of a species near the electrode compared to the concentration in the bulk electrolyte when a relatively high amount of current is extracted from the battery. This concentration difference, which is associated with limitations in the mass transfer of reactant or product species, can cause an electrical potential loss in operation of a battery. Thus, there should be an upper limit for current that can be extracted from a battery. The maximum current that can be extracted in a battery cell is called the limiting current density. The limiting current density depends on the type of electrolyte, electrode and diffusivity of interacting species. The other possible causes of a concentration gradient in the electrolyte are low diffusivity and low mobility of species that reduce the rate of electrochemical reactions. These are the main reasons for *concentration polarization* phenomenon in which the reacting and product species have difficulty in reaching the reaction sites or move away from them. In the following, various types of transport are described; then the concentration polarization resulting from this phenomenon is discussed.

3.4.1 Transport phenomena in a battery cell

Similar to applying a potential difference across a conductor to create a driving force for electron transport, applying an electric field across a solution that contains ions forms a driving force for the flow of ions. This flow of ions or *ionic current* can be described as the summation of flux of charged species in an ionic solution. That is,

$$i = \sum_i F z_i j_i \tag{3.37}$$

where i denotes the current density in an ionic solution, F Faraday's constant, z_i the charge of species i and j_i the molar flux density of species i. The movement of ions in an electrolyte can be in the forms of migration, diffusion and convection. Migration refers to the movement of ions in an electrolyte due to an imposed electrical field, diffusion is the movement of ions due to a concentration gradient (species diffuse from high concentration regions to lower concentration regions) and convection is the

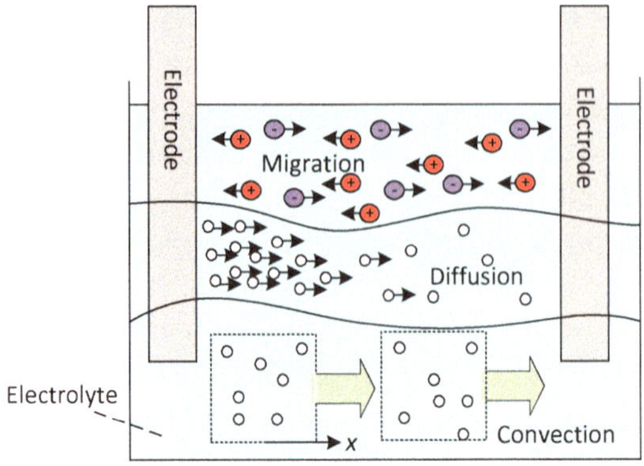

Fig. 3.5 Mass transport in a battery cell by migration, diffusion and convection.

movement of a bulk fluid (see Fig. 3.5). The net flux of ions is the summation of these three types of movements.

The molar flux density of ion i due to migration, which is called the migrational flux density, can be expressed as:

$$j_{i,Migration} = -Fz_i u_i c_i \nabla \Phi \tag{3.28}$$

where u_i is the velocity of ion i, c_i is the concentration of species i and $\nabla \Phi$ is the gradient in electrical potential. Thus, the current density due to migration can be written as:

$$i_{Migration} = -\left(F^2 \sum_i z_i^2 u_i c_i\right) \nabla \Phi \tag{3.29}$$

where the quantity in parentheses indicates the mobility of charged species, which is called ionic conductivity κ. That is, $\kappa = F^2 \sum z_i^2 u_i c_i$. As noted earlier, the flow of ions creates an ionic current in an electrolyte. This means that each ion contributes to the ionic current by carrying a fraction of the current in the solution. This fraction is called the *transference number* (*tr*). The transference number of ion i can be expressed as:

$$tr_i = \frac{Fz_i j_i}{i} = \frac{Fz_i j_i}{\sum_i Fz_i j_i} \tag{3.30}$$

Differences in the transference number of ions stem from the difference in electrical mobility of those ions. The electrical mobility of an ion reflects its ability to move rapidly across a solution in response to an electric field. The flux of an ion by migration is proportional to its transference number.

It is clear that the sum of transference numbers for all ions in a solution/electrolyte is always unity. The molar flux associated with migration can be written in terms of transference number as:

$$j_{i,Migration} = \frac{tr_i}{Fz_i} i \tag{3.31}$$

By movement of charge species toward the electrodes, a concentration gradient is created across the electrolyte. In fact, a transference number of less than one for an ion indicates that there are some fluxes from other ions in an electrolyte. These fluxes create the concentration gradient which is the driving force for diffusion phenomena. The molar flux density of ion i due to diffusion can be written in terms of the diffusion coefficient of species i (D_i) and the gradient in concentration that is caused by species i (∇c_i), as:

$$j_{i,Diffusion} = -D_i \nabla c_i \tag{3.32}$$

Thus, the current density due to diffusion can be written as follows:

$$i_{Diffusion} = -\sum_i Fz_i D_i \nabla c_i \tag{3.33}$$

It should be noted that the transference number is defined as the portion of current that is carried through migration of an ion in the absence of a concentration gradient. This means that, when the passage of ions that are moving across the electrolyte is stopped before the formation of a concentration gradient, the transfer of ions is only due to migration (movement due to an imposed electrical field) and not diffusion. So, the higher the transference number, the lower is the concentration gradient. It is noteworthy that high transference numbers mitigate the accumulation of ions (concentration polarization) in the vicinity of an electrode surface during charge or discharge.

As noted previously, the movement of a bulk solute is referred to convection. The molar flux density of ion i due to convection can be determined based on the velocity of the bulk fluid (v) and the concentration of ion i as:

$$j_{i,Convection} = c_i v \tag{3.34}$$

Thus, the current density associated with convection can be written as:

$$i_{Convection} = -\sum_i Fz_i c_i v \tag{3.35}$$

Note that, since the solution is electrically neutral, the summation of all concentrations of anions and cations is zero (i.e., $\sum c_i z_i = 0$). Thus, the convection transport process in a neutral solution does not yield ionic current. However, the movement of a bulk solution results in mixing of the solution and subsequently affects the concentration distribution in the solution. For example, the movement of a bulk solution through force or free convection can direct the reacting ions effectively toward the surface of the electrode.

The total current density in an electrolyte is the summation of current densities due to migration, diffusion and convection. That is,

$$i = -\left(F^2 \sum_i z_i^2 u_i c_i\right) \nabla \Phi - \sum_i F z_i D_i \nabla c_i - \sum_i F z_i c_i v \tag{3.36}$$

The transport phenomena in an electrolyte can be characterized by several transport properties: electrical conductivity (κ), diffusion coefficient of ions and transference number of ions. In addition, the diffusion coefficient and transference number of ions determines the magnitude of the concentration and the electrical conductivity determines the internal (ohmic) resistance. The diffusion coefficient for a solution at ambient temperature is typically on the order of 10^{-5} cm^2/s.

When the electrolyte is electrically neutral and no concentration gradient is present, the last two terms in Eq. (3.36) are zero and the expression reduces to Ohm's law, which is based on the ionic conductivity, as follows:

$$i = -\left(F^2 \sum_i z_i^2 u_i c_i\right) \nabla \Phi = -\kappa \nabla \Phi \tag{3.37}$$

When there is a concentration gradient within an electrically neutral electrolyte, the gradient in the electrical potential corresponds to

$$\nabla \Phi = -\frac{i}{\kappa} - \frac{F}{\kappa} \sum_i z_i D_i \nabla c_i \tag{3.38}$$

This equation indicates that, even in the absence of current density (i.e., $i = 0$), there would be an electrical potential in the electrolyte due to the *diffusion potential* [i.e., $(F/\kappa)\sum_i z_i D_i \nabla c_i$] caused by concentration potential. In a solution across which a concentration gradient is present, the potential difference between any two points is due to the concentration difference. This potential difference that is due to the concentration gradient is referred to as the *concentration overpotential*. The concentration overpotential occurs in a battery cell when there is a concentration difference between the reactants and products in the bulk electrolyte and on the surface of the electrode. Recall Fig. 3.1, where the concentration overpotential is seen to be equal to the potential difference between two reference electrodes, one located near the working electrode and the other located in the bulk region. Eq. (3.38) can be modified in the form of diffusion potential and integrated from the electrode surface to the bulk region to yield the concentration overpotential. That is,

$$\eta_c = E_{conc} + \int_0^{bulk} \left(\frac{1}{\mathrm{K}} - \frac{1}{\mathrm{K}_{bulk}} \right) i_x dx \tag{3.39}$$

where η_c is the concentration overpotential and x is the distance from the bulk electrolyte. The first and second terms on the right side of this equation account for the diffusion potential and overpotential associated with the variation of conductivity with concentration. The term E_{conc} denotes the potential difference across the nonuniform concentration solution (the electrolyte in a battery). This potential difference accounts for the sole effect of concentration difference between two points of a solution that are separated by a porous medium to allow diffusion transport only (see Fig. 3.6). In fact, E_{conc} corresponds to the potential that is applied to prevent the flow of current (migration transport) between two electrodes that are

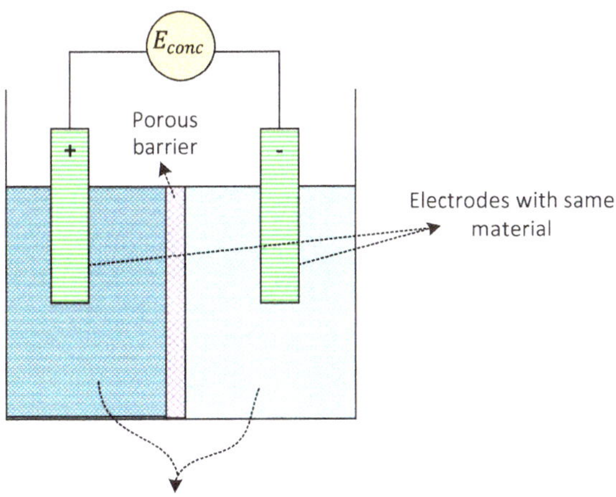

Fig. 3.6 A concentration cell in which the potential difference is only due to the concentration difference between two electrolytes having the same composition.

different only in concentrations of adjacent solutions. The following equation can be used for E_{conc}:

$$E_{conc} = \frac{RT}{F}\sum_i \ln\frac{\left(c_i^{\nu_i}\gamma_i^{\nu_i}\right)_{bulk}}{\left(c_i^{\nu_i}\gamma_i^{\nu_i}\right)_0} + \frac{RT}{F}\int_0^{bulk}\sum_j \frac{tr_j}{z_j}\frac{\partial \ln c_j\gamma_{j,n}}{\partial x}dx \quad (3.40)$$

where γ_i is the activity coefficient of species i, and j is phase indicator. Note that the first term in this equation corresponds to the concentration-dependent term in the Nernst equation (Eq. 2.32), which can be used to determine the equilibrium potential. On comparing Eq. (3.40) with the Nernst equation, it can be seen that the equilibrium potential depends on the concentration of products and reactants at equilibrium, while Eq. (3.40) in addition accounts for the diffusion potential that is caused by a concentration difference in a nonuniform solution. In fact, the terms in Eq. (3.40) demonstrate the irreversibility due to the diffusion potential caused by a concentration gradient in phase j and junction regions (boundary between the phases).

3.5 Ohmic overpotential

The last term in Eq. (3.3) (i.e., η_{ohm}) represents the overpotential caused by resistance to the flow of current. The electrical resistance is due to the passage of electronic current and ionic current through a battery cell. The electronic current is associated with the passage of electrons that are emitted from the anode through the interface between the components of the cell and also the external circuit. The passage of this electronic current leads to the loss of voltage and the loss magnitude is related to the extent of electrical current that is extracted from or inserted to the cell. The ionic current, on the other hand, is associated with the passage of ions and neutral species through the electrolyte. The resistance of the electrolyte to the flow of species leads to another loss in the voltage of a battery cell. For a battery cell that is designed properly, the main source of ohmic loss is often the ionic resistance of the electrolyte.

The ohmic loss associated with the passage of electronic current can be written as:

$$\eta_{ohm,e} = -\frac{I}{\sigma} \quad (3.41)$$

where $\eta_{ohm,e}$ is the ohmic overpotential associated with the electronic current, I is the total current drawn from the cell, σ is the electronic conductivity of electrode.

Recall Eq. (3.29), in which the flow of ions in a uniform electrolyte under the influence of an electric field was given by:

$$i = -\kappa \nabla \Phi \tag{3.42}$$

where κ is the ionic conductivity. This equation can be used to evaluate the ionic resistance in a battery cell. The ionic resistance depends on the geometry of a battery cell. For instance, for a cylindrical battery in which the current flows in the radial direction only (see Fig. 3.7), Eq. (3.42) can be written as:

$$i(r) = -\kappa \frac{d\Phi}{dr} \tag{3.43}$$

The relation between the current density i and total current I that is extracted (or inserted) from this cylindrical battery can be expressed in terms of the height of battery (H) as:

$$i(r) = \frac{I}{2\pi r H} \tag{3.44}$$

Substituting Eq. (3.44) into Eq. (3.43) yields the total loss of potential in the battery cell due to the passage of ionic current. That is,

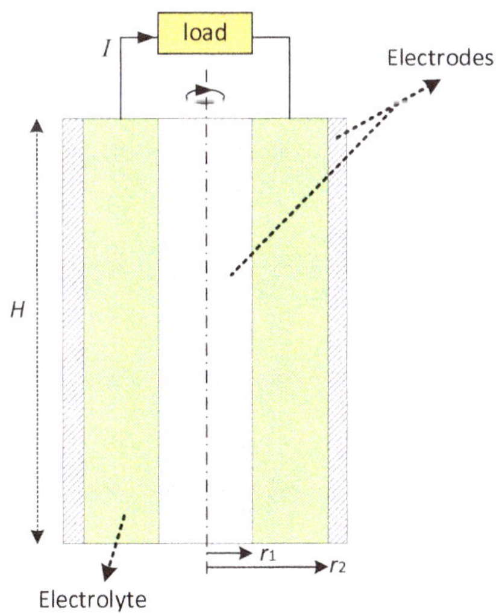

Fig. 3.7 Vertical cross section of an upright cylindrical battery.

$$\Phi(r_2) - \Phi(r_1) = \eta_{ohm,i} = -\frac{I}{2\pi r H \kappa} \ln \frac{r_2}{r_1} \tag{3.45}$$

where $\eta_{ohm,i}$ is ohmic overpotential associated with the ionic flow resistance. In this equation, the ohmic overpotential is evaluated for a constant ionic conductivity (κ) of the electrolyte. However, the effect of change in ionic conductivity with concentration has been accounted for in the concentration overpotential expression (Eq. 3.39). As can be seen in Eq. (3.45), in addition to the cell geometry, the ionic conductivity influences the ohmic overpotential. In fact, the ionic resistance is lower at higher ionic conductivities. Therefore, selection of a proper electrolyte is important as it provides for effective transport of ions, which results in lower potential loss in a battery cell.

3.6 Performance of a battery cell

The electrochemical model of a battery cell is described in this chapter. Fig. 3.8 shows briefly the electrical potential and reactant concentration profiles in a half-cell of a battery including an electrode and a surrounding

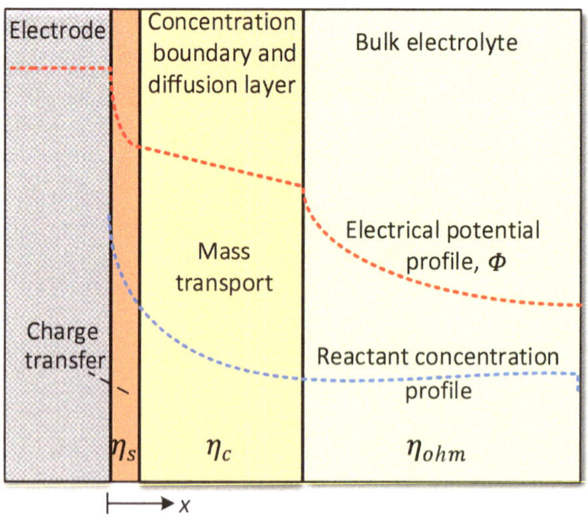

Fig. 3.8 Electrical potential and reactant concentration profiles and overpotential regions in a battery half-cell including an electrode and a surrounding conductive electrolyte.

conductive electrolyte. In addition, the regions in the half-cell in which the surface, concentration and ohmic overpotentials (η_s, η_c, and η_{ohm}) occur are shown in Fig. 3.8.

The performance of a battery cell can be evaluated by defining various performance indicators. One commonly used performance indicator is energy efficiency, which compares the battery network that is delivered during discharge or required during charge at real conditions with the same quantity at the ideal condition. It was pointed out earlier that, at equilibrium conditions where there are no irreversibilities due to overpotentials, the cell potential is referred to as the equilibrium (or reversible) potential (E_{eq}), which corresponds to the open circuit potential (E_{ocp}). In fact, the open circuit potential shows how the electrochemical interactions in that situation are near to the ideal (reversible) conditions.

Fig. 3.9 shows a typical electrical potential-capacity (amount of charge that a battery can deliver during discharge) curve for a battery cell during discharge. This graph shows the dependence of battery electrical potential (voltage) on the amount of charge that is used during discharge. At the beginning of the discharge period (to the far left of Fig. 3.9), the electrical potential of the battery is at its maximum value, which is the open circuit potential. At some point (e.g., point A), the battery cell voltage begins to drop rapidly, mainly due to the large reduction in exchange current density, which leads to a significant increase in the surface overpotential. So, near point A, the concentration polarization increases substantially. The rate at which the battery cell voltage declines depends on the types of materials used for construction of battery electrodes and the electrolyte. It is seen in Fig. 3.9 that the ohmic, surface and concentration overpotentials reduce the potential of a battery from the open circuit potential to the actual potential. The

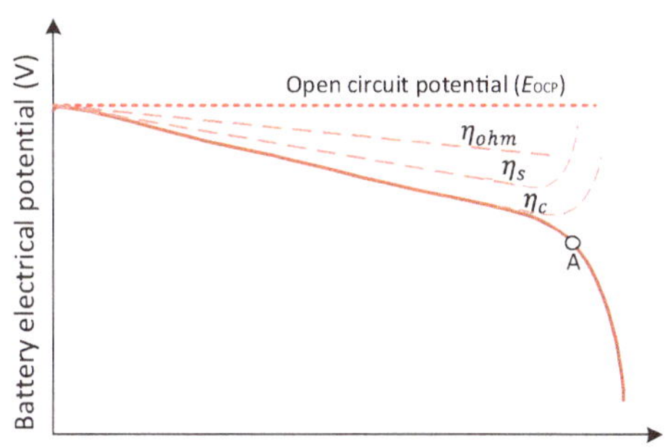

Fig. 3.9 Typical electrical potential-capacity relation for a battery cell during discharge.

ohmic overpotential increases linearly with electrical current during discharge.

As a consequence, in ideal conditions where the battery operates reversibly, the cell potential is equal to the open circuit potential. Consequently, the work output from the battery is a maximum during discharge and the work input is a minimum during charge. The energy efficiency (η_{energy}) of an actual battery during discharge can be defined as the ratio of the network output from the battery and the reversible work. That is,

$$\eta_{energy,discharge} = \frac{\int IEdt}{\int IE_{ocp}dt} = \frac{\int \left(IE_{ocp} - \eta_s - \eta_c - \eta_{ohm}\right)dt}{\int IE_{ocp}dt} \tag{3.46}$$

Similarly, during charge mode, the energy efficiency of a battery can be expressed as the reversible work input to the battery to the actual work input to the battery:

$$\eta_{energy,charge} = \frac{\int IE_{ocp}dt}{\int IEdt} = \frac{\int IE_{ocp}dt}{\int \left(IE_{ocp} - \eta_s - \eta_c - \eta_{ohm}\right)dt} \tag{3.47}$$

It is pointed out that the electrochemical study of a battery can be carried out with reasonable accuracy in one-dimension through the thickness of the cell (including the electrolyte and electrodes thicknesses), since this thickness is much smaller than the two other dimensions of the cell. Fig. 3.10 shows a prismatic battery system in which each cell in the battery contains three layers: anode and cathodes and a separating material between them. The thicknesses of the layers in a single cell are on the order of tens of micrometers, whereas the width and height of a cell are on the order of tens of centimeters. So, employing the one-dimensional electrochemical model across the thickness of the cell results in minor errors compared to employing a three-dimensional model.

In recent years, there have been numerous studies on the electrochemical analysis of various types of battery systems. Through these studies, the thermal behavior of the cell in terms of heat generation and the temperature effect on the electrical performance of the battery can be determined. For example, the thermal behavior of lithium-ion batteries is comprehensively discussed by Galatro et al. (2020). In addition, Dees et al. (2002) performed electrochemical modeling of lithium polymer batteries. Li et al. (2019)

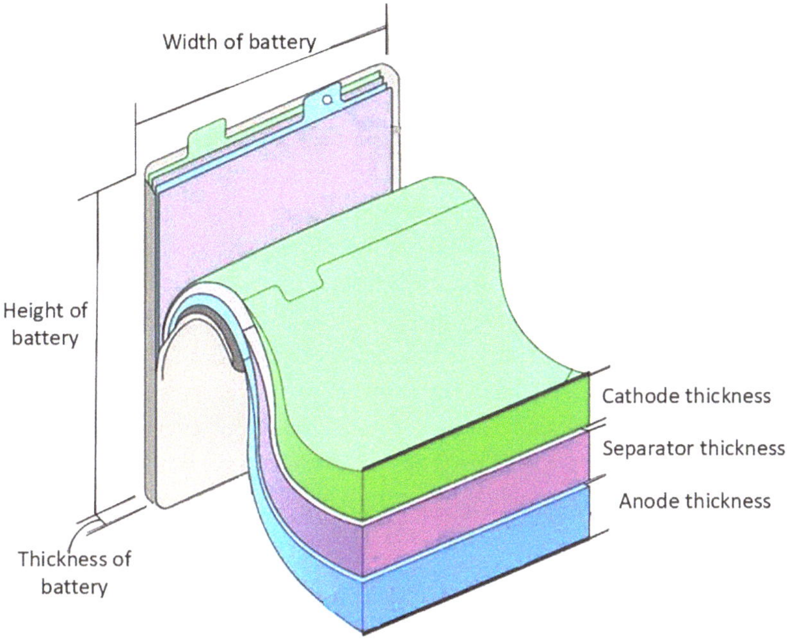

Fig. 3.10 Dimensions for a prismatic battery. *(Modified from Al-Zareer, M., 2019. Development and modeling of novel battery thermal management systems for electric and hybrid electric vehicles. PhD thesis, Ontario Tech University, Oshawa, Canada.)*

performed a similar study for a lithium-ion battery considering nonuniform concentration distribution effects. Also, Al-Zareer et al. (2017, 2019, 2020) conducted electrochemical analyses and assessments of thermal performance management systems for lithium-ion batteries. The thermal conductivity of cylindrical lithium-ion batteries are determined by Al-Zareer et al. (2021) using an electrochemical model to determine the battery operating potential and heat generation rate. Yin et al. (2019) proposed a new charging method based on an electrochemical model for lithium-ion batteries in which the side reactions are also considered. Mei et al. (2019) studied the effect of design parameters on battery performance. They also performed an optimization based on the electrochemical-thermal coupling model. An artificial neural networks approach was applied for the design of lithium-ion batteries by Wu et al. (2018). Many performance assessments of renewable energy-based battery systems have also been reported (Taslimi et al., 2021; Cai et al., 2020; Rosen and Farsi, 2022; Dehghan Abnavi et al., 2019; Zhang et al., 2018).

3.7 Closing remarks

In this chapter, an electrochemical model of a battery cell is presented and described. The electrochemical model allows the determination of the sources of overpotentials due to irreversibilities in the electrochemical reaction, as well as charge and mass transport processes. The surface overpotential is described thoroughly, explaining the concepts of potential loss during electrochemical reactions. In addition, the concentration overpotential is described, clarifying the effects of diffusion in the potential loss. The ohmic loss is explained as the electronic and ionic resistances to the flow of electrons and ions, respectively. It is shown how the overpotentials of a battery reduce the cell potential. Finally, the performance of a battery system is defined based on how nearly it approaches the performance at reversible conditions. The electrochemical study of batteries can provide insight for designing and developing suitable materials for batteries.

Study questions

3.1. What is the difference between the potential, equilibrium potential and overpotential in a battery cell?

3.2. Define the surface overpotential in a battery cell. List and describe the main reason(s) for the presence of this overpotential.

3.3. Define the concentration overpotential in a battery cell. List and describe the main reason(s) for the presence of this overpotential.

3.4. Define the ohmic overpotential in a battery cell. List and describe the main reason(s) for the presence of this overpotential.

3.5. What are the three means of mass transport in a battery cell? Explain them.

3.6. How does the reactant concentration and electrical potential vary in a battery cell when moving from the positive or negative electrode to the bulk electrolyte?

3.7. Describe and explain the energy efficiency of a battery cell. How is the energy efficiency related to the irreversibilities in a battery cell?

3.8. How does the energy of an electron change in the oxidation and reduction reactions occurring in a battery cell? If a positive potential is applied to the anode, will the energy of electron increase or decrease? Why?

3.9. What is the activation energy of a battery cell? How is it related to the Butler-Volmer equation?

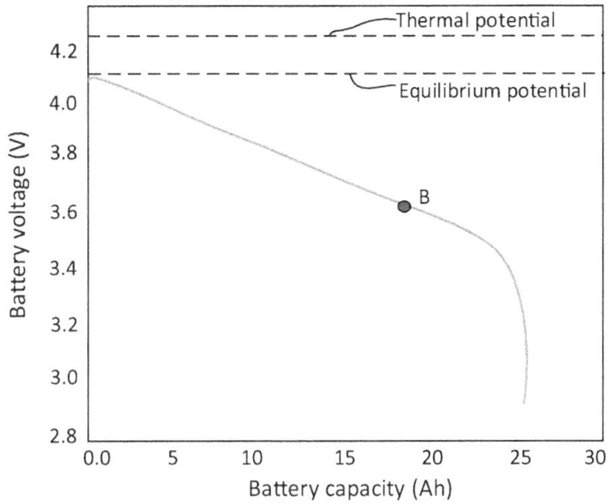

Fig. 3.11 Electrical potential-capacity curve for a lithium ion battery during discharge.

3.10. Consider the electrical potential-capacity curve shown in Fig. 3.11 for a lithium ion battery during discharge. The discharge time is 1 h. Determine the following at design point B:

- the actual electrical power that is provided by the lithium ion battery.
- the ideal electrical power (i.e., no overpotential in the battery cell) that can be provided by the lithium ion battery.
- the energy efficiency of the lithium ion battery during discharge.

3.11. In Study Question 3.10, if it were possible to somehow convert the entire energy released by the electrochemical reaction in the lithium ion battery to electrical energy, an output electrical potential of 4.3 V could be attained. This potential is called the thermal potential and it is shown in Fig. 3.11. Based on data provided in Study Question 3.10, at point B (shown in Fig. 3.11) calculate the heat generation rate in the lithium ion battery during discharge.

References

Al-Zareer, M., Dincer, I., Rosen, M.A., 2017. Electrochemical modeling and performance evaluation of a new ammonia-based battery thermal management system for electric and hybrid electric vehicles. Electrochim. Acta 247, 171–182.

Al-Zareer, M., Dincer, I., Rosen, M.A., 2019. A novel approach for performance improvement of liquid to vapor based battery cooling systems. Energy Convers. Manag. 187, 191–204.

Al-Zareer, M., Dincer, I., Rosen, M.A., 2020. A thermal performance management system for lithium-ion battery packs. Appl. Therm. Eng. 165, 114378.

Al-Zareer, M., Michalak, A., Da Silva, C., Amon, C.H., 2021. Predicting specific heat capacity and directional thermal conductivities of cylindrical lithium-ion batteries: a combined experimental and simulation framework. Appl. Therm. Eng. 182, 116075.

Cai, W., Li, X., Maleki, A., Pourfayaz, F., Rosen, M.A., Alhuyi Nazarie, M., Bui, D.T., 2020. Optimal sizing and location based on economic parameters for an off-grid application of a hybrid system with photovoltaic, battery and diesel technology. Energy 201, 117480.

Dees, D.W., Battaglia, V.S., Bélanger, A., 2002. Electrochemical modeling of lithium polymer batteries. J. Power Sources 110 (2), 310–320.

Dehghan Abnavi, M., Mohammadshafie, N., Rosen, M.A., Dabbaghian, A., Fazelpour, F., 2019. Techno-economic feasibility analysis of stand-alone hybrid wind/photovoltaic/diesel/battery system for the electrification of remote rural areas: case study Persian Gulf coast-Iran. Environ. Prog. Sustain. Energy 38 (5), 13172.

Farsi, A., Zamfirescu, C., Dincer, I., Naterer, G.F., 2020. Electrochemical transport in CuCl/HCl (aq) electrolyzer cells and stack of the Cu-Cl cycle. J. Electrochem. Soc. 167 (4), 044515.

Galatro, D., Al-Zareer, M., Da Silva, C., Romero, D., Amon, C., 2020. Thermal behavior of lithium-ion batteries: aging, heat generation, thermal management and failure. Front. Heat Mass Transfer 14, 17.

Li, J., Wang, D., Pecht, M., 2019. An electrochemical model for high C-rate conditions in lithium-ion batteries. J. Power Sources 436, 226885.

Mei, W., Chen, H., Sun, J., Wang, Q., 2019. The effect of electrode design parameters on battery performance and optimization of electrode thickness based on the electrochemical-thermal coupling model. Sustain. Energy Fuels 3 (1), 148–165.

Rosen, M., Farsi, A., 2022. Sustainable Energy Technologies for Seawater Desalination. Elsevier, Academic Press, United Kingdom.

Taslimi, M., Ahmadi, P., Ashjaee, M., Rosen, M.A., 2021. Design and mixed integer linear programming optimization of a solar/battery based conex for remote areas and various climate zones. Sustain. Energy Technol. Assess. 45, 101104.

Wu, B., Han, S., Shin, K.G., Lu, W., 2018. Application of artificial neural networks in design of lithium-ion batteries. J. Power Sources 395, 128–136.

Yin, Y., Hu, Y., Choe, S.Y., Cho, H., Joe, W.T., 2019. New fast charging method of lithium-ion batteries based on a reduced order electrochemical model considering side reaction. J. Power Sources 423, 367–379.

Zhang, W., Maleki, A., Rosen, M.A., Liu, J., 2018. Optimization with a simulated annealing algorithm of a hybrid system for renewable energy including battery and hydrogen storage. Energy 163, 191–207.

CHAPTER 4

Thermal behavior of batteries

Chapter objectives

- Describe the thermal behavior of battery systems.
- Explain the connection between the thermal behavior of a battery system and heat generation, aging mechanisms, thermal failure, and thermal management systems.
- Develop a thermal-electrochemical coupled model of battery cells.
- Determine the average cell temperature and the temperature distribution inside a battery cell using the thermal-electrochemical coupled model.
- Describe and discuss challenges and opportunities regarding the impact of thermal issues on the performance and lifetime of batteries.
- Describe and discuss the primary degradation mechanisms of cathode and anodes.
- Describe and discuss the impact of temperature, rate of discharge/charge and state of charge on aging for various battery cell chemistries.

Nomenclature

a	surface area per unit volume of porous electrode (m^2/m^3)
C_p	specific heat at constant pressure (J/kg K)
DOD	depth of discharge (%)
E	cell potential (V)
E_{act}	activation energy (J/mol)
E_{eq}	equilibrium potential (V)
E_H	enthalpy potential (V)
$\overline{H}$	species partial molar enthalpy (J/mol)
h_{conv}	convective heat transfer coefficient (W/cm^2 K)
i	current density (A/m^2)
i_n	interfacial current density (A/m^2)
I	cell current (A)
J	molar flux of species (mol/cm^2 s)
k	thermal conductivity (J/cm s K)
P	pressure (Pa)
$\dot{Q}$	heat rate (W)
$\dot{q}$	heat flux (W/cm^2)
R	universal gas constant (8.3143 J/mol K)
R_j	homogeneous production rate of species j (mol/cm^3 s)
S	entropy (kJ/K)
SOC	state of charge (%)
T	temperature (K, °C)
T_a	ambient temperature (K, °C)

Battery Technology
https://doi.org/10.1016/B978-0-443-18862-6.00005-7

t	time (s)
v	velocity (m/s)
$V_{\mathbf{cell}}$	cell volume (m^3)
dv	differential volume element of battery system (m^3)
x	direction indicator

Greek letters

v	volume (m^3)
η	overpotential (V)
Φ	electrical potential (V)
ρ	density (kg/m^3)
τ	stress (N/cm^2)
φ	general temperature-dependent property
δ	cell thickness (μm)

Subscripts

a	anode
c	cathode
gen	generation
j	phase index
l	reaction number index
s	source
ref	reference
1	solid phase
2	electrolyte phase

Acronyms

LCO	lithium cobalt oxide
LFP	lithium iron phosphate
LMO	lithium manganese oxide
LTO	lithium titanate oxide
NCA	nickel cobalt aluminum oxide
NMC	nickel manganese cobalt
SEI	solid electrolyte interface

4.1 Introduction

Heat generation inside a battery cell is due to the resistance to electrochemical reactions and movement of species within the cell. The generation of heat can be analyzed through thermodynamic and electrochemical examinations of battery systems, as are thoroughly described in Chapters 2 and 3. This generated heat is transported by conduction through several materials and interfaces to reach the outer surface of a battery, where convection to an external cooling fluid takes place to dissipate heat from the battery. So, the operation and storage temperature of a battery both play a key role in the lifetime, performance, safety and ultimately cost of a battery system. The thermal behavior of batteries necessitates the application of thermal management systems for battery systems. This chapter describes the thermal behavior of battery systems and its connection to heat generation, aging mechanism, thermal failure, and thermal management systems, by examining the major thermal issues within battery systems. The thermal-electrochemical coupled model is also presented, with which the average cell temperature and the temperature distribution inside a cell can be determined. In addition, challenges and opportunities are discussed in relation to the impact of thermal issues on the performance and lifetime of batteries.

4.2 Aging mechanism in batteries

Aging in battery systems generally refers to the degradations that are caused by chemical and physical interactions in the electrolyte, the electrodes, and the separator. Aging in a battery system presents itself through a decrease in delivery of power and capacity (known as power and capacity fade) and an increase in the internal resistance. More specifically, aging in a battery system prevents the ion exchange process from delivering its full capacity, which results in a decrease in battery performance. This performance decrease is typically determined based on the change in capacity over time and the battery state of health.

The battery *state of health* is an indicator for the present battery capacity compared to the capacity at the beginning of its life. This indicator can be used to predict the end of life of a battery. For example, the end of life for a lithium-ion battery that is used in an electric vehicle is about 80% of the nominal capacity at the beginning of its life. The state of health for a battery is not a physical characteristic itself, but instead is related to several parameters such as internal resistance, decrease in capacity and power and number of charge-discharge cycles. When the amount of charge that a battery can store starts to decrease with repeated use, it experiences a *fade of capacity or capacity loss*. Similarly, *power fade* occurs in a battery when the amount of charge that can be delivered at the rated voltage decreases with use due to an increase in the internal resistance of the battery. Various approaches have been reported to estimate the state of health and aging in battery systems. These include the following:

- Equivalent circuit models (Birkl et al., 2017; Eddahech et al., (2015)). In these models, the battery is reduced to an equivalent circuit model.
- Electrochemical models (Doyle and Newman, 1995; Newman and Tiedemann, 1975; Afshar, (2017); Dalverny et al., 2011; Wagemaker et al., 2011). In these models, the phenomena occurring in a battery are each modeled.
- Correlations of stress factors with capacity fade (Galatro et al., 2020b; Rohr et al., 2017; Wu et al., 2019). In these methods, the rise in impedance can be determined through accelerated aging tests.

Fig. 4.1 shows the capacity fade profile and relative internal resistance of a lithium-ion battery over time at various operating temperatures. The internal resistance of the battery is measured with respect to the relative value at the initial conditions (Keil, 2017). At a specific lifetime of 0.3 years (about 100 days), it is seen that as the operating temperature increases from −10°C to 50°C, the capacity fade increases from 2% to 7%. In addition, the internal resistance of the battery rises over time, and this effect becomes more pronounced as the battery temperature increases. It is therefore clear that the temperature has a significant impact on aging mechanisms in batteries. The operating temperature for most batteries is usually recommended to be between 20°C and 40°C, as this typically allows better performance and a longer lifetime for the battery. This desired temperature range of battery can be achieved through the use of battery thermal management systems.

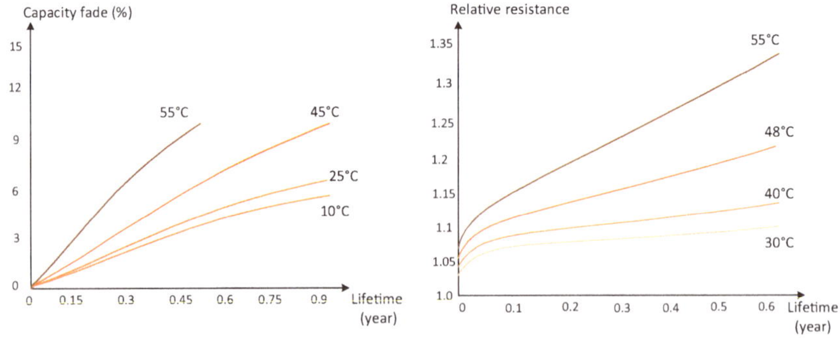

Fig. 4.1 Variation with lifetime of capacity fade (Schimpe et al., 2018) and relative internal resistance (Keil, 2017) for several operating temperatures for a lithium-ion battery.

The thermal behavior of a battery is dependent on its intrinsic features such as intrinsic thermal stability, heterogeneity and nonuniformity on the surface cell temperatures and extrinsic factors such as *state of charge*, current load and operating temperature. The extrinsic factors are referred to extrinsic stress factors as they increase the degradation processes in a battery system that lead to a decrease in its performance over time. The state of charge (SOC) represents the ratio of available capacity at a given time to the maximum possible charge that can be stored in battery. Thus, a value of SOC of 100% indicates a fully charged battery and of 0% indicates a fully discharged battery. The SOC of a battery can be measured based on the properties of the active electrode materials. For a lithium-ion battery, for example, when all cyclable lithium ions are transferred to the anode, the state of charge is 100% while at a 0% state of charge, all of these ions are transferred back to the cathode. In addition, the quality of the structure of the electrodes determines the percentage of active materials that can yield cyclable ions.

4.2.1 Cycling aging and calendar aging

Aging can occur through a *cycling mode* and a *calendar mode*. Aging in the *calendar mode* refers to the irreversible loss of capacity when a battery is stored or in a rest state. At rest, the battery is exposed to the environment temperature without any external effect which might accelerate the rate of aging. The calendar aging can be intensified or weakened by storage conditions like temperature and state of charge. In lithium-ion batteries, for instance, relatively high temperatures (over 40°C) and relatively low temperatures (lower than 10°C) result in the loss of active ions and subsequently lower ionic interactions and diffusions, which lead to capacity fade for the battery. As another stress factor for calendar aging of a battery, SOC can have another effect. For identical type of batteries at the same temperature but different SOC levels, deterioration occurs at different rates (Barré et al., 2013). The higher the level of SOC (especially over 70%), the higher is the rate of deterioration; this is due to undesirable side reactions that occur at high SOC levels (Palacín, 2018).

Aging in the cycling mode refers to irreversible losses during charging and discharging cycles. Like calendar aging, which is made worse by an increase in temperature and SOC, cycling aging is also made more pronounced by these two stress factors. In addition, the rate of discharge/charge (C-rate) and variation of SOC are other important factors that can influence aging mechanisms during charge/discharge. Cycling aging is increased at higher discharge/charge rates and higher variations in SOC (Santhanagopalan et al., 2014). The charging/discharging voltages during the life of a battery are other factors impacting the aging process. A high charging voltage increases the degradation rate in batteries (Kötz et al., 2010).

In addition to the already noted stress factors for calendar and cycling aging, cell chemistry also can have an important effect. In terms of calendar aging, for example, nickel manganese cobalt (NMC), lithium cobalt oxide (LCO), and lithium manganese oxide (LMO) batteries are more sensitive to aging in high-temperature operations than nickel cobalt aluminum oxide (NCA) and lithium iron phosphate (LFP) batteries (Galatro et al., 2020a, b). Also, in cycling aging, LFP and NMC batteries are more sensitive to aging at high C-rates (fast charging) compared to NCA batteries (Galatro et al., 2020a,b). There have been numerous studies on aging during cycling and calendar modes and on the effect of various factors on the rate of these aging mechanisms. Table 4.1 summarizes several recent studies for aging during cycling and calendar modes for various cell chemistries and operating conditions.

4.2.2 Aging in lithium-ion batteries

There have been many studies on aging mechanism in batteries, especially for lithium-ion cells. For example, Birkl et al. (2017) examined the cause and effect of aging mechanisms in lithium-ion batteries. Fig. 4.2 shows the cause and effect analysis of degradation mechanisms in lithium-ion batteries.

In lithium-ion batteries, the formation of a solid electrolyte interface (SEI) can occur. This process can be described as follows: At the start of the first charging process, lithium ions are de-intercalated from the cathode to the electrolyte and then migrate to the anode. As a result of these interactions, boundary phases of the SEI are created at interfaces that enable a reversible cycling of the battery. These interfaces are formed on electrolyte-electrode interfaces on the cathode and anode and they are more pronounced at the anode side (see Fig. 4.3). The SEI is typically made of insoluble decomposition products of electrolyte and lithium ions from the cathode. However, the composition varies with temperature. In addition, the SEI provides a passivation layer on the surface of the anode that prevents further decomposition of the electrolyte. The growth in thickness of the SEI results in an increase in internal resistance during intercalation and de-intercalation of lithium ions. Thus, the performance of the battery at various operating temperatures depends on the stability and passivation effect of the SEI. Therefore, the use of battery thermal management systems is

Table 4.1 Summary of key findings on aging mechanisms during cycling and calendar modes for different cell chemistries, based on selected studies.

Aging mode	Investigators	Key findings
Calendar mode	Keil et al. (2016), Ecker et al. (2012), Gismero et al. (2019), Dubarry et al. (2018), Eddahech et al. (2015)	• For all cell chemistries, capacity fade significantly increases with storage temperature, whereas SOC has a lesser impact on degradation mechanisms in batteries • At high temperature operations, NMC, LCO, and LMO cells are more sensitive to aging than NCA and LFP cells • At low temperature operations, rapid capacity fade is reported for LFP batteries • At low and high storage temperatures, lithium titanate oxide (LTO) batteries are reported to be less sensitive to aging compared to LCO, NMC, NCA, and LFP batteries
Cycling mode	Han et al. (2014), Wang et al. (2011), Wu et al. (2017), Devie et al. (2014), Petzl and Danzer (2014), Tippmann et al. (2014), Leng et al. (2015)	• For all cell chemistries, the rate of charging (C-rate) and temperature strongly increase aging during cycling mode at high temperatures ($T > 40°C$) and low temperatures ($T < 20°C$) • At high C-rates, LTO and NCA batteries are more suitable than LFP and NMC batteries by being resistant to cycling aging

necessary to maintain the battery temperature at the proper range to assure the ability for passivation and the stability of the SEI.

In lithium-ion batteries, aging in the anode is the result of degradations caused by temperature, SOC and overcharge stress factors. At relatively high temperatures (above 50°C) and high SOC values (above 80%), the solid

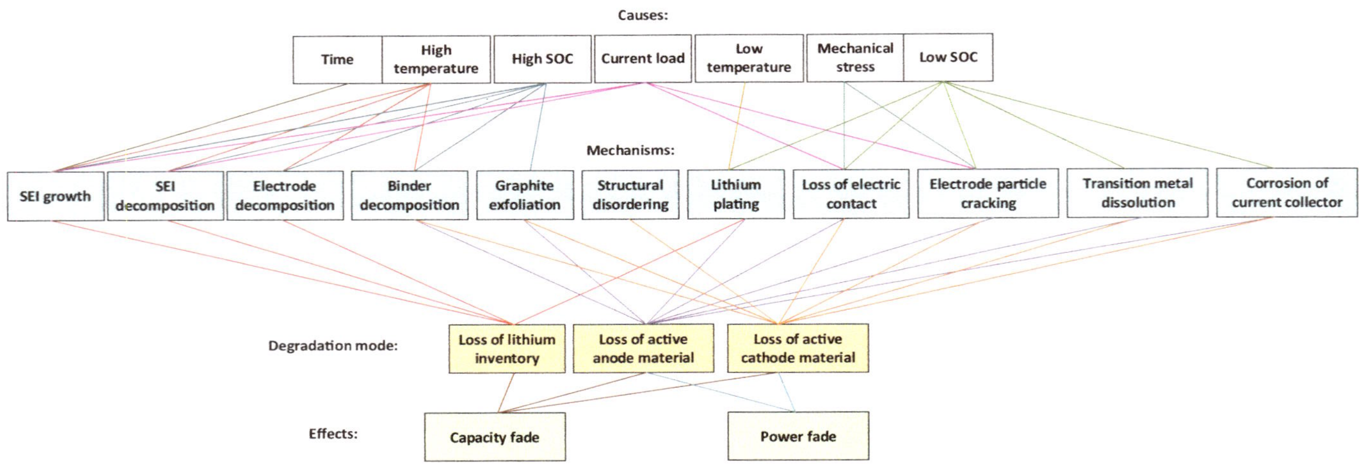

Fig. 4.2 Cause and effect analysis of degradation mechanisms in lithium ion batteries, as reported by Birkl et al. (2017).

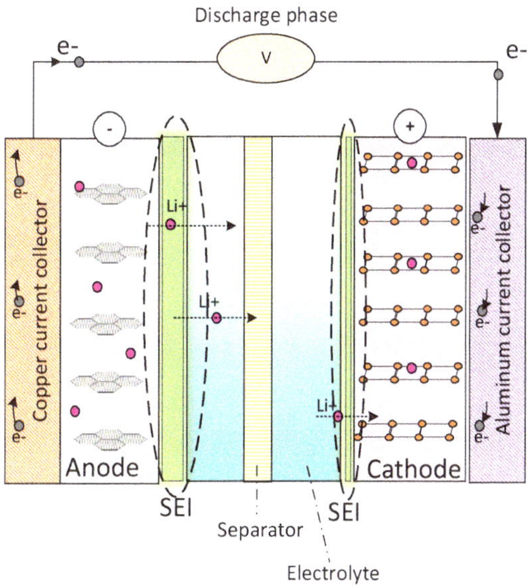

Fig. 4.3 Passage of lithium ions from the solid-electrolyte interfaces (SEIs) in a lithium-ion battery.

electrolyte interface gradually dissolves. Less permeable species like lithium salts are formed from it. These lithium salts form a resistance to movement of lithium ions and subsequently increases the resistance in the anode. In addition, the diffusion of lithium into the SEI and anode declines at relatively low temperatures (below 20°C). This results in the deposition of lithium on the anode in the form of dendrites. This process is referred to as *lithium plating* (see Fig. 4.4). The consequences of lithium plating are worse than those of the high temperature operation of the battery as it increases the aging mechanism in the battery. Fig. 4.4 illustrates lithium deposition (lithium plating) and SEI decomposition during low-temperature aging.

The aging mechanism in the cathode is due to the effects of temperature and SOC. Fig. 4.5 shows the degradation mechanisms in the cathode and their effects on the material and performance of a lithium-ion battery. As noted earlier, the SEI is also formed at the cathode but it is thinner than at the anode. So, the effect of SEI thickness and passivation ability is not as prominent at the cathode as at the anode. In addition, the chemistry of the cell, like the material composition of the anode and cathode, has a significant effect on the performance and lifetime of a battery. The development of novel materials for the anode can potentially improve the performance and energy density capability of batteries.

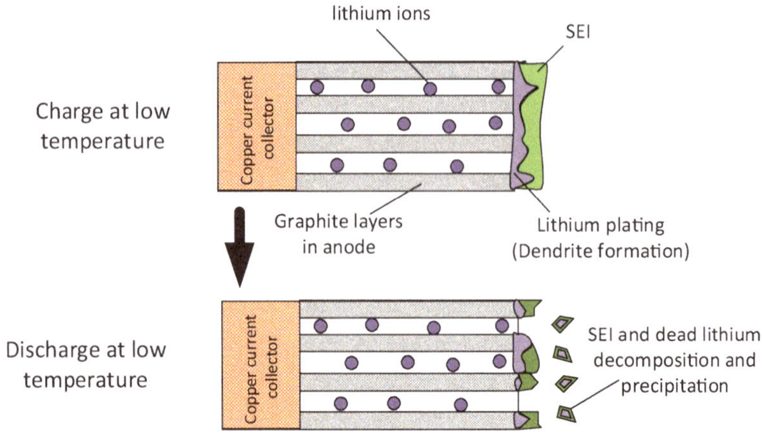

Fig. 4.4 Lithium plating and SEI decomposition in anode of a lithium-ion battery.

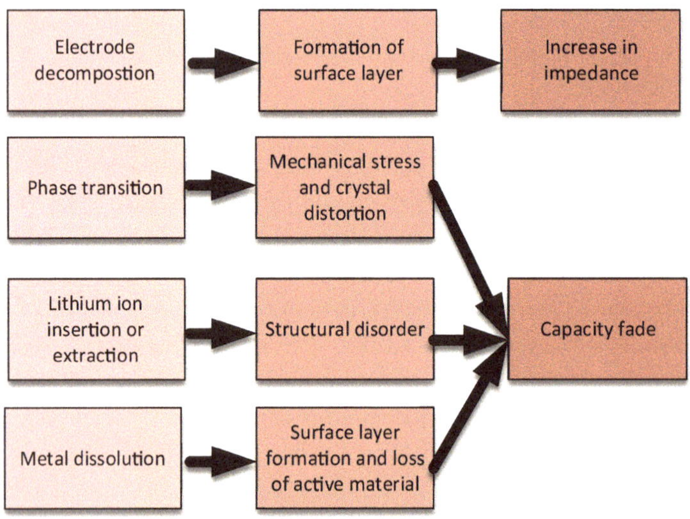

Fig. 4.5 Aging mechanisms in cathode of lithium-ion battery, and their effects on materials and on battery performance. *(Data extracted from Lin, C., Tang, A., Mu, H., Wang, W., Wang, C., 2015. Aging mechanisms of electrode materials in lithium-ion batteries for electric vehicles. J. Chem. 2015, 11.)*

4.3 Thermal runaway

In the operation of battery systems, some extreme events may happen that cannot be controlled, even in the presence of battery thermal management systems. Such events often lead to cell failure. In some instances, a large

Table 4.2 Onset temperatures for selected electrode chemistries in lithium-ion batteries.

Electrode	Chemistry	Approximate onset temperature (°C)
Anode	Artificial graphite	130
	Ordinary graphite	120
Cathode	Nickel cobalt aluminum oxide (NCA)	300
	Lithium manganese oxide (LMO)	300
	Lithium cobalt oxide (LCO)	250–270
	Lithium iron phosphate (LFP)	250
	Nickel manganese cobalt (NMC)	120

Data from Golubkov, A.W., Scheikl, S., Planteu, R., Voitic, G., Wiltsche, H., Stangl, C., Fauler, G., Thaler, A., Hacker, V., 2015. Thermal runaway of commercial 18650 Li-ion batteries with LFP and NCA cathodes – impact of state of charge and overcharge. RSC Adv. 5(70), 57171–57186; Kvasha, A., Gutiérrez, C., Osa, U., de Meatza, I., Blazquez, J.A., Macicior, H., Urdampilleta, I., 2018. A comparative study of thermal runaway of commercial lithium ion cells. Energy 159, 547–557.

amount of heat can be generated in a single cell and this heat may cause a thermal runaway of other cells in the vicinity of the cell failure, leading to failure of the entire battery pack.

Thermal runaway in a battery happens when the temperature of the cell exceeds the *onset temperature*, the temperature at which battery self-heating commences due to decomposition of the SEI. With further increases in cell temperature, the degradation reactions occur at a faster rate, eventually emitting various gases and leading to ignition and combustion of the battery cell. The conditions that lead to thermal runaway can be internal short circuits between electrodes, between current collectors and between the anode and the collector. Table 4.2 lists the onset temperatures for several electrode chemistries in lithium-ion batteries.

Thermal runaway in batteries can be also due to overdischarging or overcharging, which generates more heat than normal discharging or charging rates. The excess generated heat cannot be dissipated from the battery, resulting in an increase in the battery temperature. With this increase in temperature, the chemical and electrochemical interactions within the cells occur at a faster rate, resulting in even more heat generation that cannot be dissipated. The possible exothermic reactions in a battery are discussed elsewhere (Spotnitz and Franklin, 2003). This chain of interactions (see Fig. 4.6) in the battery is difficult to stop once started. The continuation of the thermal runaway process leads to the failure and destruction of the battery, usually through the stored energy in battery being released suddenly by fire and/or explosion. Furthermore, the temperature rise in a single cell

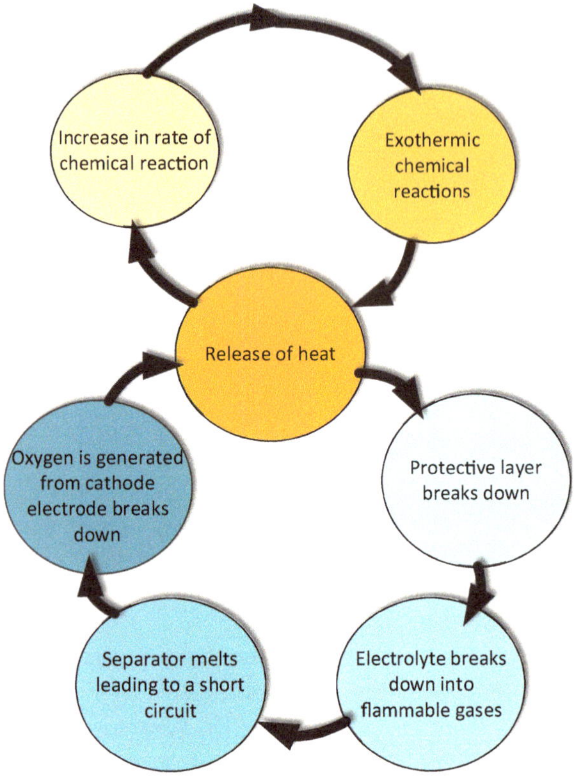

Fig. 4.6 Thermal runaway process in a lithium-ion battery system.

during thermal runaway affects other cells in the vicinity, creating a domino effect. So, the failure and destruction of one cell often leads to the destruction of the entire battery stack and subsequently the battery pack.

There are also other factors that can prompt thermal runaway like external/internal short circuits, overcharge/overdischarge and overheating. The internal short circuit can be caused by impurities present in the battery due to issues with the manufacturing process or a crash involving the battery. Fig. 4.6 shows the steps for thermal runaway in a lithium-ion battery.

To prevent thermal runaway, the current and voltage of charging should be limited, as that permits the temperature of the battery to be controlled. In addition, appropriate cooling systems should be employed to maintain the battery temperature in the desired temperature range (around the environment temperature). As mentioned before, the thermal runaway in one part of battery pack can affect the other parts, so the best way to stop this phenomenon is to remove the battery from the facility.

4.4 Heat generation rate and temperature variation in batteries

Identifying different sources of heat generation in batteries is important as it allows for employing appropriate strategies to mitigate these sources of heat and subsequently prolonging the life of the batteries. In this section, sources of heat generation in a battery are briefly introduced (recalling they are comprehensively described in Chapters 2 and 3). Also, the role of stress factors that can intensify the heat generation rate in a battery are discussed.

Heat is produced in a battery from three fundamental sources: (i) resistance to flow of electrons (i.e., Joule or ohmic heating), (ii) resistance to electrochemical reactions (i.e., heat generation during charge transfer between electrode and electrolyte in electrode reactions), and (iii) intercalation and de-intercalation of active ions (i.e., entropic heating), which changes the atomic configurations in the electrode. Joule heating is known as polarization heating (or irreversible heating), whereas entropic heat is reversible heat as it indicates the heat generated during the reversible isothermal operation of a battery.

A general expression for the heat generation rate $\dot{Q}_{gen}$ in a cell can be written based on the thermodynamic energy balance for the cell at a uniform temperature, as follows:

$$\dot{Q}_{gen} = I(E_{eq} - E) - I\left(T\frac{\partial E_{eq}}{\partial T}\right) \tag{4.1}$$

where I denotes cell current, E_{eq} equilibrium potential, E cell potential, and T absolute temperature of the cell. The first term on the right side denotes the *irreversible* heat generation rate, including charge and mass transfer overpotentials and ohmic losses in a battery cell. The second term states the *reversible* or entropic heat rate in a battery cell. Phase change and mixing effects are neglected in this equation. When the anode and cathode temperatures are identical, the entropic (or reversible) heat generation rate can be determined based on the thermodynamic entropy change rate within the cell (i.e., TdS/dt). Note that the electrode potential in Eq. (4.1) is determined based on the *average concentration of all species in the cell*. The use of the average concentration may result in a significant error in the estimation of heat generation rate. Thus, Rao and Newman (1997) modified the expression in Eq. (4.1) to consider the *average local concentration* in each phase and then applied Faraday's law to calculate the heat generation rate for a battery cell.

The local heat generation rate method, accounting for local thermal effects in the battery, leads to the following:

$$\dot{Q}_{\text{gen}} = -\int_{v}\left(\sum_{l} a i_{n,l} E_{H,l} dv\right) - IE \qquad (4.2)$$

where a is the surface area per unit volume of a porous electrode, $i_{n,\ l}$ is the interfacial current density for reaction l, and $E_{H,\ l}$ is the enthalpy potential that corresponds to the reaction enthalpy change. The reversible heat effect can be determined by taking the equilibrium potential (E_{eq}) to mean enthalpy potential (E_H). The first term on the right side of Eq. (4.2) states the average enthalpy potential (also entailing the entropic heat) that is integrated across the battery thickness where the electrochemical reaction takes place.

The *local heat generation rate* in a battery is developed later by Rao and Newman (1997) and Thomas and Newman (2003). It accounts for heat generation at the interfaces and bulk materials in different regions of the cell. The thermal effect associated with a concentration gradient through the electrolyte is neglected in their analysis. For *one dimensional analysis* of the *local irreversible heat generation flux* (i.e., irreversible heat generation rate per unit superficial area of cell), the integration is carried out over the thickness of electrode (see Fig. 4.7). That is,

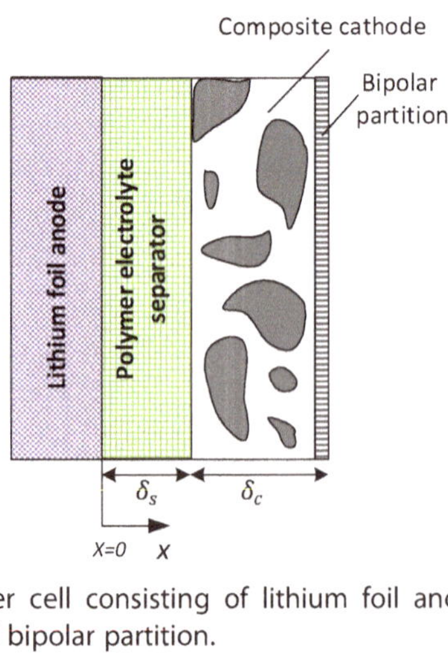

Fig. 4.7 Lithium-polymer cell consisting of lithium foil anode, polymer electrolyte, composite cathode, and bipolar partition.

$$\dot{q}_{\text{gen}} = i\Delta\Phi + i\eta_a + \int_{x_1}^{x_2} \left(ai_n\eta_c - i_1\frac{d\Phi_1}{dx} - i_2\frac{d\Phi_2}{dx} \right) dx \tag{4.3}$$

This equation presents the total irreversible heat generation including Joule (or ohmic) heating in the electrolyte and separator and resistance to anodic and cathodic reactions (i.e., surface overpotential in both anode and cathodes). The first term on the right side (i.e., $i \Delta \Phi$) is the ohmic heating flux, which corresponds to the potential difference between two phases of the separator (i.e., $\Delta \Phi = \Phi(0) - \Phi(x_1)$). The term $i\eta_a$ denotes the heat generation flux in the anode, and $\int_{x_1}^{x_2}(ai_n\eta_c - i_1 d\Phi_1/dx - i_2 d\Phi_2/dx)dx$ denotes the heat generation flux in the cathode, electrolyte and electronically conducting matrix (i.e., solid phase), respectively. It has also been found (Rao and Newman, 1997; Thomas and Newman, 2003) that the results obtained from neglecting concentration gradient in the electrolyte are the same, assuming no mixing effects. The mixing effect is accounted for later by Thomas and Newman (2003), through a Taylor-series applied for the enthalpy of species in the electrolyte.

In Table 4.3, the main factors are presented that affect the reversible and irreversible heat generation rates in a battery cell. The effects of these factors on a battery are made evident by variations in the cell temperature. So, by measuring the cell temperature, the heat generation rate in a battery can be determined indirectly. Furthermore, an intrinsic heterogeneity in material and physical properties of a battery may results in an uneven cell temperature distribution. So, fabrication of uniform materials in battery cells and proper cooling and heating of battery packs via thermal management systems are important for even distribution of battery surface temperature.

4.5 Thermal behavior model of batteries

The thermal behavior model of battery systems can be electrochemically and thermally coupled or decoupled depending on assumptions considered for estimation of the heat generation rate in the battery systems. As mentioned previously, the operation of a battery system depends on the cell temperature and rate of charge/discharge. The decoupled models use empirical correlations to describe the charge/discharge rate curves (obtained from experiments) with the assumption of constant cell temperature. However the decoupled model is simpler than the coupled model, and it is accurate when the battery is not sensitive to temperature. *Thermal-electrochemical* coupled models first produce the potential and current from the model then use this

Table 4.3 Several factors affecting reversible and irreversible heat generation in a battery cell.

Factor	Effects on heat generation in battery
Electrode microstructure[a]	Lower porosity of electrode results in an increase in ohmic heat in the electrolyte and more hotspots in the vicinity of the electrode separator interface. The electrode microstructure influences the distribution of temperature and heat generation in a battery cell
State of charge (SOC)[a]	Entropic heat increases at low SOC (<25%) and high depth of discharge (DOD > 75%), which results in increased reversible heat. Note: DOD is the percentage of capacity that has been removed from the fully charged battery
Ambient temperature (T_0)[a]	Low ambient temperatures and higher rates of discharge result in increased irreversible heat generation due to increased polarization losses
Charge and discharge C-rate[a]	Higher charge and discharge rates (>1C) lead to higher heat generation rates, as ohmic loss increases at high C-rates. Also, low C-rates (<1C) result in an increase in reversible heat generation
Battery size[b]	An increase in battery size causes an uneven distribution of active materials in the electrode, which results in an uneven distribution of heat generation rate (specifically ohmic heat rate) in a battery cell
Cell chemistry[c]	Reversible heat is more significant in LCO batteries, while in LMO, NCM, and LFP cells, the irreversible heat increases with capacity. Also, heat generation rates are in almost the same range for LFP, LMO, and LCO batteries with the same graphite anode, while heat generation rates in NMC batteries are higher compared to other cell chemistries

[a]Kantharaj and Marconnet (2019).
[b]Santhanagopalan et al. (2014).
[c]Kantharaj and Marconnet (2019), Lin et al. (2017).

information to determine the heat generation rate inside the battery and subsequently to predict the temperature distribution within the cell. This, in turn, is used to produce the new cell potential and current.

Pals and Newman (1995) proposed a partially-coupled model to reduce the complexity of coupled models. In their model, they assumed that the heat generation rate inside a battery is almost independent of the discharge history of the battery.

The general thermal energy equation for a multicomponent system like battery, which is derived from a differential energy balance, can be written as:

$$\rho_j C_{p_j}\left(\frac{\partial T_j}{\partial t} + v_j \cdot \nabla T_j\right) + \left(\frac{\partial \ln \rho_j}{\partial \ln T_j}\right)\left(\frac{\partial P_j}{\partial t} + v_j \cdot \nabla P_j\right)$$
$$= -\nabla \cdot \dot{q}_j - \tau_j \cdot \nabla v_j + \sum_{\text{species}} \overline{H}_j\left(\nabla \cdot J_j - R_j\right) \quad (4.4)$$

where subscript j denotes phase; C_p, ρ, v, and P respectively are specific heat at constant pressure, density, velocity, and pressure; $\dot{q}$ and τ respectively are heat flux and stress, $\overline{H}$ and J respectively are species partial molar enthalpy and molar flux of species due to migration and diffusion, and R_j is the homogeneous production rate of species j. In Eq. (4.4), the last term on the right side denotes the transport of thermal energy through migration and diffusion of species, with the summation accounting for all species in phase j. In addition, the last term on the left and the second and third terms on the right denote reversible and irreversible conversion of mechanical energy to thermal energy. With negligible heat effects due to viscous dissipation (i.e., $\tau \cdot \nabla v$) and negligible pressure work and body force (i.e., $(\partial \ln \rho / \partial \ln T)(\partial P/\partial t + v \cdot \nabla P)$) and no homogeneous chemical reaction (i.e., R_j), Eq. (4.4) reduces to:

$$\rho_j C_{p_j}\left(\frac{\partial T_j}{\partial t} + v_j \cdot \nabla T_j\right) = -\nabla \cdot \dot{q}_j + \sum_{\text{species}} \overline{H}_j\left(\nabla \cdot J_j\right) \quad (4.5)$$

The heat flux term ($\dot{q}$) encompasses the heat flux carried by interdiffusion of species, conduction and the Dafur heat flux. The Dafur heat flux ($\dot{q}_D$) states the diffusion-thermo effect, which often can be neglected compared to the two other heat flux terms. Thus, the heat flux in phase j can be written as:

$$\dot{q}_j = \sum \overline{H}_j\left(\nabla \cdot J_j\right) - k_j \nabla T_j + \dot{q}_D \quad (4.6)$$

where k_j is the thermal conductivity of phase j. Assuming local thermal equilibrium in the battery system, applying the continuity equation (i.e., $\nabla v_j = 0$) and neglecting the enthalpy of mixing and phase change in Eq. (4.5) leads to the following:

$$\frac{\partial\left(\rho c_p T\right)}{\partial t} + \nabla(v \cdot T) = \nabla \cdot k \nabla T + \dot{q}_s \quad (4.7)$$

This differential equation describes the temperature distribution in a battery system. The first and second terms on the left side represent the accumulation and convection of energy in the system. The first and second terms on the right side denote conduction and the heat source in the system.

We may now couple the thermal modeling described above with the electrochemical model for a battery system. This means that the physiochemical properties of a cell (e.g., ionic conductivity and diffusion coefficient of the electrolyte) that depend on the temperature are linked to the thermal model and electrochemical model. Coupling of two models can be achieved by the Arrhenius equation, which relates a general temperature-dependent property φ (e.g., the ionic conductivity) with the activation energy for the evolution process of that property ($E_{\text{act},\,\varphi}$), as follows:

$$\varphi = \varphi_{\text{ref}} \exp\left[\frac{E_{\text{act},\varphi}}{R}\left(\frac{1}{T_{\text{ref}}} - \frac{1}{T}\right)\right] \tag{4.8}$$

where φ_{ref} is the general temperature-dependent property value at reference temperature (T_{ref}). Fig. 4.8 illustrates the thermal-electrochemical coupled model. First, from the electrochemical model, the local heat generation rate due to Joule heating and chemical reactions is determined. Then, this information is employed in the conservation of thermal energy equation, from which the cell temperature can be determined. The temperature obtained

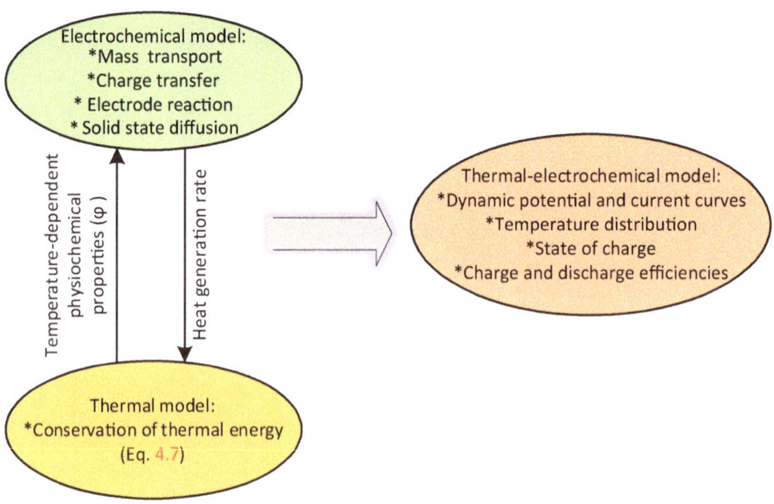

Fig. 4.8 Illustration of thermal-electrochemical coupled model approach for a battery system.

is fed back to update the calculation of physiochemical properties in the electrochemical model.

We now apply Eq. (4.7), which describes the accumulation, convection, conduction, and heat source terms, to a battery cell. For a stationary battery cell in which the bulk electrolyte is not flowing, the convection term in Eq. (4.7) can be neglected. Therefore, Eq. (4.7) can be written for transient conduction as follows:

$$\frac{\partial(\rho c_p T)}{\partial t} = \nabla \cdot k \nabla T + \dot{q}_s \tag{4.9}$$

For a single cell, which has a small thickness, the *lumped-parameter approach* is applicable. This indicates the condition where the ratio of convective heat transfer coefficient (h_{conv}) times the thickness of the cell (δ) to the thermal conductivity of the cell (k) is very small (i.e., $h_{\text{conv}}\delta/k \ll 1$). Hence, the assumption of uniform temperature throughout the cell is accurate. The conduction term is also eliminated as the temperature does not change spatially. Thus, Eq. (4.9) can be expressed based on heat removal rate per unit volume of cell to the surrounding ($\dot{q}_{\text{loss}}$) and volume averaged heat generation rate ($\dot{q}_{\text{gen}}$), as follows:

$$\dot{q}_{\text{gen}} = \frac{\partial(\rho c_p T)}{\partial t} + \dot{q}_{\text{loss}} \tag{4.10}$$

This time-dependent temperature differential equation is commonly used for lithium-polymer, lead-acid, nickel-hydrogen, and lithium-ion batteries. The heat loss rate to the surroundings usually occurs thorough convection, as:

$$\dot{q}_{\text{loss}} = \frac{1}{V_{\text{cell}}}\left(h_{\text{conv}} A_{\text{cell}}(T - T_a)\right) \tag{4.11}$$

where V_{cell} is the cell volume, h_{conv} is the convective heat transfer coefficient, A is the surface area through which heat is removed from the cell, and T_a is the ambient temperature. The heat generation rate in a battery cell ($V_{\text{cell}}\dot{q}_{\text{gen}}$) can be determined using Eq. (4.2). Substituting Eqs. (4.2) and (4.11) into Eq. (4.10) and modifying to convert the unit volume to the total volume of the cell, yields the following expression for the heat generation rate in a battery cell, as follows:

$$-\int_v \left(\sum_l a i_{n,l} U_{H,l} dv\right) - IE = \frac{\partial(m c_p T)}{\partial t} + h_{\text{conv}} A_{\text{cell}}(T - T_a) \tag{4.12}$$

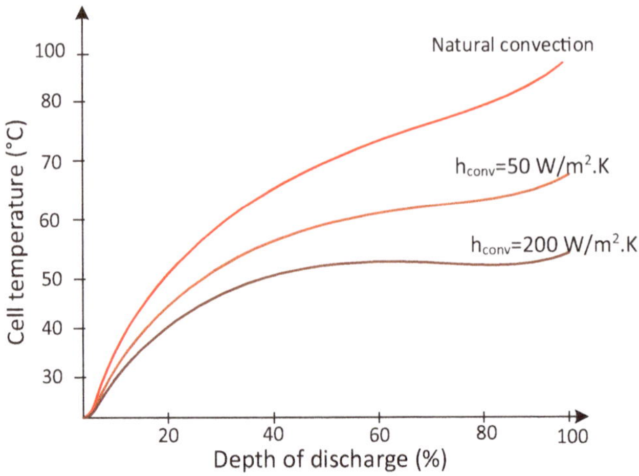

Fig. 4.9 Variation of cell temperature with charge input during 3C charging at several convective heat transfer coefficients for a lithium-ion battery. *(Data from Chen, S.C., Wan, C.C., Wang, Y.Y., 2005. Thermal analysis of lithium-ion batteries. J. Power Sources 140(1), 111–124.)*

This equation is similar to the heat generation rate equation obtained from an energy balance for a battery cell (i.e., Eq. 2.47) which is described in Chapter 2. Fig. 4.9 shows the variation of temperature with charge input during 1C charging for several convective heat transfer coefficients for a lithium-ion battery. It is seen that the temperature of the cell increases as the depth of discharge approaches 100%. Furthermore, at a higher heat convection rate with the surroundings (i.e., a higher value for the convective heat transfer coefficient), a higher heat rate is dissipated from the cell and subsequently a lower cell temperature is achieved.

4.6 Thermal behavior impacts in batteries: Challenges and opportunities

In order to make battery systems more widely used and competitive, especially in vehicle applications where the main alternative is the traditional combustion engine, battery users need to be able to recharge their batteries anywhere and quickly through *fast charging*. Although the required time for charging a battery system is reduced by fast charging methods, the rate of heat generation in the battery is increased significantly in this process, which leads to higher temperature gradients and increases in the heterogeneity of the battery in a manner that favors the SEI growth. In addition, when fast charging is applied at low temperature ranges (less than −20°C), the phenomenon of lithium plating occurs, with lithium plating formed in the battery and resulting in a notable capacity loss. Thus, the application of battery thermal management systems is necessary to provide a cooling effect when using fast charging and to provide heating at very low cell temperatures (cold weather conditions) to avoid lithium plating.

The *challenges and opportunities* associated with the effect of thermal behavior and temperature of a battery on its lifetime and performance are mainly focused on practices for delaying aging and strategies for minimizing the impact of fast charging on cell heterogeneity and temperature gradient. In the following, the main strategies and practices are described for decelerating the aging mechanism and reducing or minimizing the impact of fast charging on cell heterogeneity and temperature gradient.

4.6.1 Minimizing the impact of stress factors on battery aging mechanisms

Several strategies can be employed to reduce the rate of aging in battery systems to lower or minimize the effect of stress factors on the degradation mechanism of batteries. These include:

(i) the use of thermal management systems,
(ii) the development of novel materials for the anode, and
(iii) the practices of battery users including following instructions on using and maintaining a battery.

Proper *practice of battery users* delays both cycling aging and calendar aging in batteries. This suggests to have a fully charged battery at room temperature rather than at high temperatures, and to leave a battery which is at rest at room temperature rather than at low temperatures. This also suggests avoiding a high level of depth of discharge (percentage of discharge relative to the capacity of the battery) (Wikner and Thiringer, 2018).

The proper design of *battery thermal management systems* enhances battery performance and lifetime by maintaining the operating temperature at a desired range and at a uniform temperature throughout the battery. In addition, thermal management systems prevent thermal runaway in batteries by dissipating the generated heat in the battery. Numerous studies have been reported for various types of battery thermal management systems (Al-Zareer et al., 2018, 2019a,b, 2020). These are comprehensively discussed in Chapter 5. The design of battery thermal management systems is based on the thermal characteristics of the battery cell, module and pack considering temperature variations in a single cell and series or parallel integrations of the cells. During cycling aging, battery thermal management systems should maintain the cell temperature within the desired range to ensure the stability of the SEI in the battery cell. Also, battery thermal management systems should prevent different aging paths in the battery by providing an even temperature distribution among the cells. Finally, battery thermal management systems should control the temperature rise during high rates of charging (i.e., fast charging).

Many areas related to the development of *new anode materials* or modifications of currently used anode materials for battery systems have been the subject of investigations and assessments. For lithium-ion batteries, for example, Mao et al. (2018) found that stable sizes for graphite crystallite significantly enhance the lifetime of a battery. In addition, Mussa et al. (2019) showed that the application of nanocomposites like cobalt oxide allows for high-temperature (up to 100°C) operation of lithium-ion batteries with stable electrochemical performance.

There have been many studies to find the most suitable anode material, instead of using graphite anodes, to diminish aging mechanisms in lithium-ion batteries (e.g., eliminate lithium plating and SEI growth) and to minimize intrinsic heterogeneity in batteries (created during fabrication of anode), which intensifies the temperature gradient between the cells. Graphene and graphene-based transition to graphene oxide have been identified as the most appropriate materials that can be used instead of graphite for the anode. Luo et al. (2018) demonstrated that utilizing a graphene oxide like metal sulfides in addition to elemental graphene in the anode of a lithium ion battery substantially decreases heat generation inside the battery and subsequently improves the performance of the lithium-ion battery. Although, graphene has a relatively higher capacity and lower rate of aging, it has a low initial coulomb efficiency (ratio of the ions participating in the electrochemical reactions to the total ions). Therefore, further studies are still required to identify proper materials for the anode that exhibit high coulomb efficiency, as well as high capacity, low temperature rise during fast charging and low cost.

4.6.2 Fast charging methods

There are several methods for fast charging. These include:

(i) combined constant potential and constant current charging,
(ii) pulse charging, and
(iii) negative pulse charging.

In the first method, a constant current is first applied by a charger until the battery potential approaches a defined potential. Above this potential, the voltage is maintained at a constant value and the current decreases slightly until a full charge is achieved. In the pulse charging method, pulses of current are sent to the battery which reduces the time of charging. However, the heat generation rate with this method increases considerably, so selection of a suitable pulse frequency can potentially reduce impedance and subsequently heat generation in the battery. Negative pulse charging is considered a promising charging method, as it permits cell temperature rise to be reduced or minimized by applying small discharges to the battery during the rest period of pulse charging. This leads to depolarization of the cell, allowing for high charging rates.

The fast charging method is limited by several factors:

(i) the limiting current at which the cell overpotential increases drastically, and

(ii) the kinetics of the battery, through which the extent of aging is described.

An overview for various methods for fast charging in battery systems is provided by Tomaszewska et al. (2019).

4.7 Closing remarks

In this chapter, the thermal behavior of batteries and its relation to aging mechanisms, thermal failure, heat generation, and thermal management systems are described. The thermal-electrochemical coupled model is presented and described. Through this model, the average cell temperature and temperature distribution in a cell can be determined. The primary degradation mechanisms of the cathode and anodes are discussed. As part of this, the impact of temperature and rate of discharge/charge and state of charge on aging for various cell chemistries during cycling mode and calendar mode are examined. Furthermore, sources of heat generation in a battery and the effect of stress factors on reversible and irreversible heat rates are described. It was noted that, for a low cell temperature domain ($<20°C$), the aging mechanism is mostly associated with the lithium plating while, for high cell temperature domain ($40°C < T <$ onset temperature), the degradation process is mainly due to the growth of the SEI. In an intermediate cell temperature domain, the battery lifetime and performance are at their highest values. Finally, the challenges and opportunities associated with the effects of thermal behavior of a battery on its lifetime and performance are reviewed.

The challenges are associated with strategies to mitigate aging in batteries and diminish the impact of fast charging on cell heterogeneity and the temperature gradient within a cell. The opportunities include efforts to minimize the aforementioned challenges by efficient and effective battery thermal management systems, effective fast charging methods and proper selection of the anode material.

Study questions

4.1. Define the following terms used in battery systems:
- State of health
- State of charge
- Capacity fade
- Power fade

4.2. What are the aging mechanisms in a battery system? What are two common types of aging in a battery?

4.3. How is the solid electrolyte interface (SEI) formed in a lithium-ion battery cell? What are the main factors that accelerate the SEI formation in a lithium-ion battery cell?

4.4. What are the main reasons for the phenomenon of thermal runaway in a battery system? How can thermal runaway be controlled?

4.5. List and describe the factors affecting reversible and irreversible heat generation in a battery cell.

4.6. What are the actions that can decelerate aging mechanisms in a battery system?

4.7. Consider the voltage-discharge capacity diagram for a lithium-ion battery at different C-rates (rates of discharge). It is observed that the effective capacity of a battery cell is reduced when discharge occurs at relatively high rates. The discharge time is 1 h. The temperature is uniform throughout the battery cell, which has a surface area of $100\,\text{cm}^2$ and a mass of 100 g. The equilibrium potential varies linearly with temperature based on the following equation:

$$E_{eq} = 1.43 \times 10^{-4}T^2 + 0.04T - 2.1 \quad \left(\text{for } E_{eq} \text{ in } V \text{ and } T \text{ in K}\right)$$

If the temperature at design point A shown in Fig. 4.10 is 25°C, 27°C, 30°C, and 34°C at $C = 0.1$, 1, 2, and 4, respectively:

- determine the heat generation rate in the battery at this point for C-rates of 0.1, 1, 2, and 4.

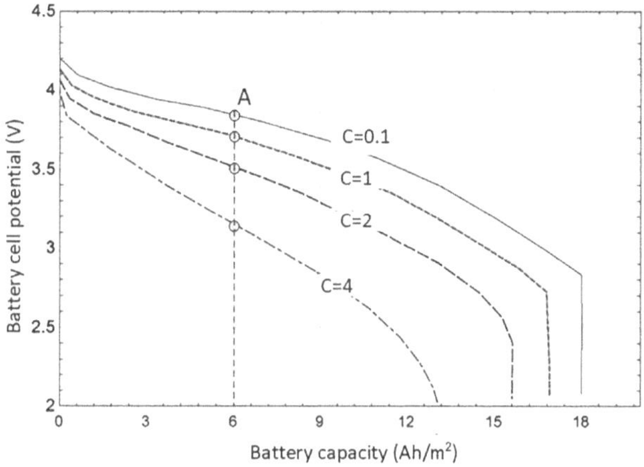

Fig. 4.10 Variation of battery cell potential with discharge capacity of the battery cell for four discharge rates (C-rates).

- discuss what should be the characteristics of a thermal management system to maintain the temperature of the battery at 25°C, 27°C, 30°C, and 34°C for C-rates of 0.1, 1, 2, and 4, respectively.

References

Afshar, S, 2017. Lithium-Ion Battery SOC Estimation, PhD Dissertation. University of Waterloo, Waterloo, Ontario, Canada.

Al-Zareer, M., Dincer, I., Rosen, M.A., 2018. A review of novel thermal management systems for batteries. Int. J. Energy Res. 42 (10), 3182–3205.

Al-Zareer, M., Dincer, I., Rosen, M.A., 2019a. Comparative assessment of new liquid-to-vapor type battery cooling systems energy. Energy 188, 116010.

Al-Zareer, M., Dincer, I., Rosen, M.A., 2019b. A novel approach for performance improvement of liquid to vapor based battery cooling systems. Energy Convers. Manage. 187, 191–204.

Al-Zareer, M., Dincer, I., Rosen, M.A., 2020. A thermal performance management system for lithium-ion battery packs. Appl. Therm. Eng. 165, 114378.

Barré, A., Deguilhem, B., Grolleau, S., Gérard, M., Suard, F., Riu, D., 2013. A review on lithium-ion battery ageing mechanisms and estimations for automotive applications. J. Power Sources 241, 680–689.

Birkl, C.R., Roberts, M.R., McTurk, E., Bruce, P.G., Howey, D.A., 2017. Degradation diagnostics for lithium ion cells. J. Power Sources 341, 373–386.

Dalverny, A.L., Filhol, J.S., Doublet, M.L., 2011. Interface electrochemistry in conversion materials for Li-ion batteries. J. Mater. Chem. 21 (27), 10134–10142.

Devie, A., Dubarry, M., Liaw, B.Y., 2014. Diagnostics of Li-ion commercial cells – experimental case studies. ECS Trans. 58 (48), 193.

Doyle, M., Newman, J., 1995. Modeling the performance of rechargeable lithium-based cells: design correlations for limiting cases. J. Power Sources 54 (1), 46–51.

Dubarry, M., Qin, N., Brooker, P., 2018. Calendar aging of commercial Li-ion cells of different chemistries—a review. Curr. Opin. Electrochem. 9, 106–113.

Ecker, M., Gerschler, J.B., Vogel, J., Käbitz, S., Hust, F., Dechent, P., Sauer, D.U., 2012. Development of a lifetime prediction model for lithium-ion batteries based on extended accelerated aging test data. J. Power Sources 215, 248–257.

Eddahech, A., Briat, O., Vinassa, J.M., 2015. Performance comparison of four lithium–ion battery technologies under calendar aging. Energy 84, 542–550.

Galatro, D., Al-Zareer, M., Da Silva, C., Romero, D., Amon, C., 2020a. Thermal behavior of lithium-ion batteries: aging, heat generation, thermal management and failure. Front. Heat Mass Transf. 14, 17.

Galatro, D., Silva, C.D., Romero, D.A., Trescases, O., Amon, C.H., 2020b. Challenges in data-based degradation models for lithium-ion batteries. Int. J. Energy Res. 44 (5), 3954–3975.

Gismero, A., Stroe, D.I., Schaltz, E., 2019. Calendar aging lifetime model for NMC-based lithium-ion batteries based on EIS measurements. In: 2019 Fourteenth International Conference on Ecological Vehicles and Renewable Energies (EVER). IEEE, pp. 1–8.

Han, X., Ouyang, M., Lu, L., Li, J., Zheng, Y., Li, Z., 2014. A comparative study of commercial lithium ion battery cycle life in electrical vehicle: aging mechanism identification. J. Power Sources 251, 38–54.

Kantharaj, R., Marconnet, A.M., 2019. Heat generation and thermal transport in lithium-ion batteries: a scale-bridging perspective. Nanoscale Microscale Thermophys. Eng. 23 (2), 128–156.

Keil, P., 2017. Aging of Lithium-Ion Batteries in Electric Vehicles. PhD Thesis, Technical University of Munich, Munich, Germany.

Keil, P., Schuster, S.F., Wilhelm, J., Travi, J., Hauser, A., Karl, R.C., Jossen, A., 2016. Calendar aging of lithium-ion batteries. J. Electrochem. Soc. 163 (9), 1872.

Kötz, R., Ruch, P.W., Cericola, D., 2010. Aging and failure mode of electrochemical double layer capacitors during accelerated constant load tests. J. Power Sources 195 (3), 923–928.

Leng, F., Tan, C.M., Pecht, M., 2015. Effect of temperature on the aging rate of Li ion battery operating above room temperature. Sci. Rep. 5 (1), 1–12.

Lin, C., Wang, F., Fan, B., Ren, S., Zhang, Y., Han, L., Liu, S., Xu, S., 2017. Comparative study on the heat generation behavior of lithium-ion batteries with different cathode materials using accelerating rate calorimetry. Energy Procedia 142, 3369–3374.

Luo, R.P., Lyu, W.Q., Wen, K.C., He, W.D., 2018. Overview of graphene as anode in lithium-ion batteries. J. Electron. Sci. Technol. 16 (1), 57–68.

Mao, C., Wood, M., David, L., An, S.J., Sheng, Y., Du, Z., Meyer III, H.M., Ruther, R.E., Wood III, D.L., 2018. Selecting the best graphite for long-life, high-energy Li-ion batteries. J. Electrochem. Soc. 165 (9), A1837.

Mussa, Y., Ahmed, F., Abuhimd, H., Arsalan, M., Alsharaeh, E., 2019. Enhanced electrochemical performance at high temperature of cobalt oxide/reduced graphene oxide nanocomposites and its application in lithium-ion batteries. Sci. Rep. 9 (1), 1–10.

Newman, J., Tiedemann, W., 1975. Porous-electrode theory with battery applications. AICHE J. 21 (1), 25–41.

Palacín, M.R., 2018. Understanding ageing in Li-ion batteries: a chemical issue. Chem. Soc. Rev. 47 (13), 4924–4933.

Pals, C.R., Newman, J., 1995. Thermal modeling of the lithium/polymer battery: I. Discharge behavior of a single cell. J. Electrochem. Soc. 142 (10), 3274.

Petzl, M, Danzer, MA, 2014. Nondestructive detection, characterization, and quantification of lithium plating in commercial lithium-ion batteries. J. Power Sources 254, 80–87.

Rao, L., Newman, J., 1997. Heat-generation rate and general energy balance for insertion battery systems. J. Electrochem. Soc. 144 (8), 2697.

Rohr, S., Müller, S., Baumann, M., Kerler, M., Ebert, F., Kaden, D., Lienkamp, M., 2017. Quantifying uncertainties in reusing lithium-ion batteries from electric vehicles. Procedia Manuf. 8, 603–610.

Santhanagopalan, S, Smith, K, Neubauer, J, Kim, GH, Pesaran, A, Keyser, M, 2014. Design and Analysis of Large Lithium-Ion Battery Systems. Artech House.
Schimpe, M., von Kuepach, M.E., Naumann, M., Hesse, H.C., Smith, K., Jossen, A., 2018. Comprehensive modeling of temperature-dependent degradation mechanisms in lithium iron phosphate batteries. J. Electrochem. Soc. 165 (2), A181.
Spotnitz, R., Franklin, J., 2003. Abuse behavior of high-power, lithium-ion cells. J. Power Sources 113 (1), 81–100.
Thomas, K.E., Newman, J., 2003. Thermal modeling of porous insertion electrodes. J. Electrochem. Soc. 150 (2), A176.
Tippmann, S., Walper, D., Balboa, L., Spier, B., Bessler, W.G., 2014. Low-temperature charging of lithium-ion cells part I: electrochemical modeling and experimental investigation of degradation behavior. J. Power Sources 252, 305–316.
Tomaszewska, A., Chu, Z., Feng, X., O'Kane, S., Liu, X., Chen, J., Ji, C., Endler, E., Li, R., Liu, L., Li, Y., 2019. Lithium-ion battery fast charging: a review. ETransportation 1, 100011.
Wagemaker, M., Singh, D.P., Borghols, W.J., Lafont, U., Haverkate, L., Peterson, V.K., Mulder, F.M., 2011. Dynamic solubility limits in nanosized olivine $LiFePO_4$. J. Am. Chem. Soc. 133 (26), 10222–10228.
Wang, J., Liu, P., Hicks-Garner, J., Sherman, E., Soukiazian, S., Verbrugge, M., Tataria, H., Musser, J., Finamore, P., 2011. Cycle-life model for graphite-$LiFePO_4$ cells. J. Power Sources 196 (8), 3942–3948.
Wikner, E, Thiringer, T, 2018. Extending battery lifetime by avoiding high SOC. Appl. Sci. 8 (10), 1825.
Wu, Y., Keil, P., Schuster, S.F., Jossen, A., 2017. Impact of temperature and discharge rate on the aging of a $LiCoO_2/LiNi_{0.8}Co_{0.15}Al_{0.05}O_2$ lithium-ion pouch cell. J. Electrochem. Soc. 164 (7), A1438.
Wu, Z., Wang, Z., Qian, C., Sun, B., Ren, Y., Feng, Q., Yang, D., 2019. Online prognostication of remaining useful life for random discharge lithium-ion batteries using a gamma process model. In: 20th International Conference on Thermal, Mechanical and Multi-Physics Simulation and Experiments in Microelectronics and Microsystems (EuroSimE). IEEE, pp. 1–6.

CHAPTER 5

Battery thermal management systems

Chapter objectives

- Introduce battery thermal management systems.
- Describe the need for thermal management systems for safe, long duration, and efficient operation of batteries, in general and specifically in high-powered applications.
- Describe and discuss conventional battery thermal management systems, including air, liquid, and phase change material-based units, and the integration between them.
- Present and describe recently developed technologies for battery thermal management systems, such as evaporative pool boiling-based battery thermal management systems.

Nomenclature

Acronyms

APM ammonium polyphosphate
EG expanded graphite
HDPL high density polyethylene
PSC primary sclerosing cholangitis

5.1 Introduction

Batteries are widely used in various devices, such as electric vehicles, smart phones, and laptop computers. Need are growing to increase battery energy capacity by stacking multiple battery cells in a pack and to charge batteries at higher currents to shorten the charging time. These actions result in higher rate of heat generation in batteries. In high-power discharging and charging applications, in particular, significant heat is generated in batteries. It is necessary to remove this heat from the battery as it adversely affects battery performance and can lead to battery failure.

The use of *battery thermal management systems* is necessary, especially in relatively high-powered applications for several reasons:

Battery Technology
https://doi.org/10.1016/B978-0-443-18862-6.00003-3

- to regulate the temperature of battery and maintain it in the optimum range, and
- to avoid or reduce any nonuniform temperature distributions within the battery, especially at high rates of discharge and/or charge and during extreme environmental conditions (e.g., extremely hot or cold conditions).

Although battery thermal management systems are usually designed to cool battery systems, they should be able to provide heat for initial operation of battery at low environment temperatures. In addition, battery thermal management systems need to provide a safe operating zone for batteries by avoiding or at least reducing the occurrence of the phenomenon of thermal runaway (i.e., allowing the battery temperature to exceed the critical point that may lead to fire and explosion). In the event of an occurrence of this phenomenon, the thermal management system needs to be able to ventilate exhaust gases from the battery.

Beyond effectiveness in managing thermal issues, various characteristics are sought in battery thermal management systems. Some of the main desired characteristics follow: lightness, compactness, low cost, easy maintenance, high reliability, easy packing, and low power consumption (Khan et al., 2017; Pesaran et al., 1999).

5.2 Design of battery thermal management systems

The design of battery thermal management systems involves numerous steps. The design procedures for battery thermal management systems are somewhat analogous to those for conventional heat exchangers. The main steps in battery thermal management system design follow:

- Identification of objectives and constraints in design of the battery thermal management system (e.g., dimensions, geometry, orientation, number, heat transfer medium, maximum pressure drop, need for ventilation, and cost).
- Evaluation of heat generation rate within battery cells and cell temperature distribution across the battery pack at various operating conditions. Such information can be attained using thermodynamic, thermal and electrochemical modeling of the batteries and battery pack.
- Determination, by measurement, analytical evaluation or estimation, of the thermophysical properties of batteries such as heat capacities of materials in components (such as the cell core or outer casing) at various states of charge and temperatures. This can be accomplished using mass-weighted averages of cell/module components and calorimeter methods (Santhanagopalan et al., 2014).
- Computation of the temperature of the module in contact with the thermal management system at various operating conditions. This is

typically carried out using energy conservation and heat transfer principles. For comprehensiveness, such work needs to consider relevant heat transfer fluids (air, liquid) as well as various flow paths and flow rates.
- Determination of the heat transfer rate between the heat transfer fluid in the battery thermal management system and the battery. This step often involves the use of computational fluid dynamics (CFD), correlations, or experimental methods.
- Evaluation of the expected performance of the battery thermal management system based on the thermal characteristics of the battery system and cooling load estimated for the thermal management system.
- Construction and testing of the primary battery thermal management system. The main system and auxiliary components like fans, pumps, evaporators and heaters, are sized and installed. Important design criteria related to energy requirements, complexity, performance and maintenance can be determined and compared with alternative systems. Practical limitations that are not accounted for in theoretical simulation are introduced as required, and if necessary designers re-do the previous design steps with the updated information so as to achieve the design specifications in the best way possible.
- Optimization of the battery thermal management system to achieve the design objectives.

As stated earlier, the performance of a battery system is highly dependent on its operating temperature, so it is important to recognize how heat is generated within a battery. To achieve this, it is important to understand battery heat generation and the impact of temperature on battery performance. Chapters 2–4 of this book describe the thermodynamic, electrochemical and thermal behavior of batteries, respectively. From these treatments, the thermal characteristics of a battery can be determined. Battery thermal management systems have been extensively investigated for various types of batteries, considering numerous designs and operating conditions.

In all designs of battery thermal management systems, it is important to understand battery system thermal behavior. Fig. 5.1 briefly describes illustratively the thermal behavior of a battery system. Heat generation in a battery is seen to originate from four sources: (i) intercalation and deintercalation of active ions (i.e., entropic heating), (ii) heat of phase change, (iii) overpotentials, and (iv) heat release due to mixing. The last three sources are known as irreversible heating whereas entropic heat is reversible heat as it represents the heat generated during the reversible isothermal operation of a battery. The heat generated due to overpotentials is

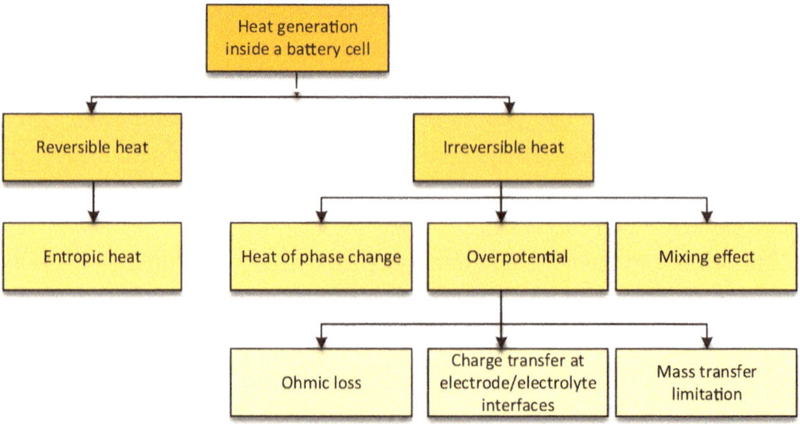

Fig. 5.1 Thermal behavior in a battery system.

seen in Fig. 5.1 to result from (i) resistance to electrochemical reactions (i.e., heat generation during charge transfer between electrode and electrolyte in electrode reactions), (ii) resistance to flow of electrons (i.e., Joule or ohmic heating), and (iii) resistance to movements of ions in the electrolyte.

5.3 Categories of battery thermal management systems

Most battery thermal management systems can be divided into three main categories:

(i) air-based
(ii) liquid-based
(iii) phase change material (PCM)-based

The latter category of battery thermal management systems includes systems based on solid-liquid phase change and liquid-vapor phase change. Conventionally, the PCM is more often used for solid-liquid phase change, so this chapter focuses on that type. In this chapter, therefore, the PCM is a substance that absorbs and releases energy by melting and solidifying, respectively, at the phase change temperature. In liquid-vapor phase change-based systems, heat pipes can be used to convey heat from the battery to a condenser via a heat transfer fluid, or batteries can be directly submerged in a stationary liquid heat transfer fluid. The former is known as a heat pipe-based battery thermal management system and the latter is known as an evaporating pool boiling-based battery thermal management system. Fig. 5.2 illustrates the main battery thermal management system categories.

Other types of battery thermal management systems have been developed based on selective integrations of the above-listed three main

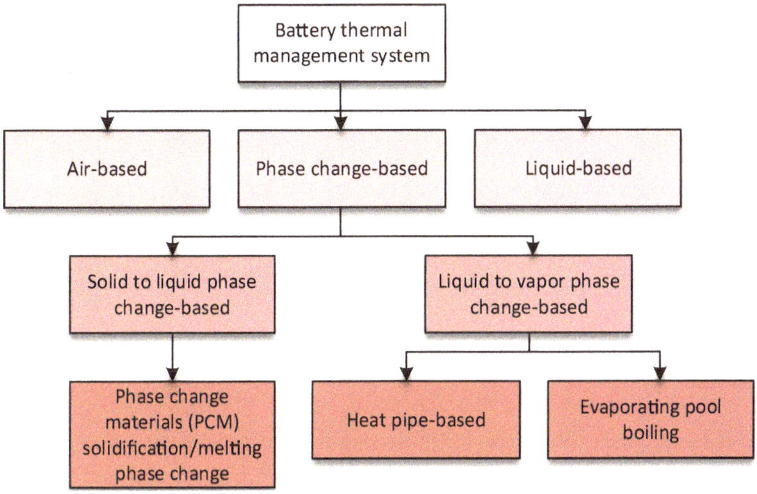

Fig. 5.2 Main categories of battery thermal management systems.

categories. For instance, a PCM-based system can be integrated with a liquid-based or air-based system. In this combined system, the PCM-based system absorbs heat from a set of batteries, and is then cooled by a liquid-based or air-based thermal management system. Often, a uniform temperature distribution can be achieved with this integrated thermal management system through the provision of uniform cooling to the battery and maintaining the temperature of battery at the desired operating condition.

Battery thermal management systems can alternatively be categorized as *passive* and *active* cooling. In the passive cooling system, the environment temperature is used to cool a battery while in the active cooling an energy source is used to provide battery cooling. The active cooling itself can be classified based on direct and indirect cooling, based on whether the cooling medium is in direct or indirect contact with a battery, respectively.

In the following sections, air-based, liquid-based, PCM-based and liquid-vapor phase change-based battery thermal management systems are described.

5.4 Air-based battery thermal management systems

Air-based thermal management systems are generally classified into *forced convection* and *natural convection* air flow systems. In the natural convection air flow, no extra energy is used to move the air into and through a battery pack, while in the forced convection air flow, energy is utilized to move the air into and through the battery pack. Forced convection air flow thermal management systems have higher heat transfer coefficients that natural convection air flow systems, resulting in more effective heat transfer between the air as a cooling medium and the battery pack. Therefore, most air-based battery thermal management systems are designed based on the forced convection air flow, so that is the focus of this section. These systems are low in cost and have simple configurations with easy maintenance. Typically, air-based battery thermal management systems are capable of providing cooling for moderate cooling load applications.

Many designs exist for the air-based battery thermal management systems. These differ in terms of path of air flow, velocity of air flow and arrangement of batteries in a pack. Selected important recent studies on air-based battery thermal management systems are now described. These are illustrative of many of the air-based battery thermal management system designs.

Fig. 5.3A and B show typical horizontal and longitudinal layouts of battery packs. In the horizontal battery pack configuration, the inlet and outlet ports for the air coolant are exposed to the larger surface areas of the cells, while in the longitudinal configuration, the inlet and outlet ports of the air flow are exposed to the thinner widths of the cells. Based on Xu and He (2013), the horizontal configuration exhibits better cooling performance than the longitudinal configuration, primarily due to the shorter air flow path. The authors also found that the horizontal configuration can be further improved by adding a bottom duct in which a higher conduction heat transfer area is provided.

Fig. 5.4 shows Z-type and U-type ducts for flow of air coolant. Sun and Dixon (2014) demonstrate that the Z-type duct performs more effectively in lowering the variations in cell temperature compared to the U-type duct. In addition, inlet and outlet port geometries of the air coolant flow impact the uniformity in the air flow rates of the cooling channels, and subsequently affect cell temperature uniformity and the pressure drop between cooling channels. Using tapered inlet and outlet ports for the flow of air can significantly reduce the variations in the air flow rate in the cooling channel and consequently can provide a more uniform cell temperature. It is also found that the pressure drop in the Z-type duct with tapered inlet and outlet ports for air flow is lower than that for the U-type duct.

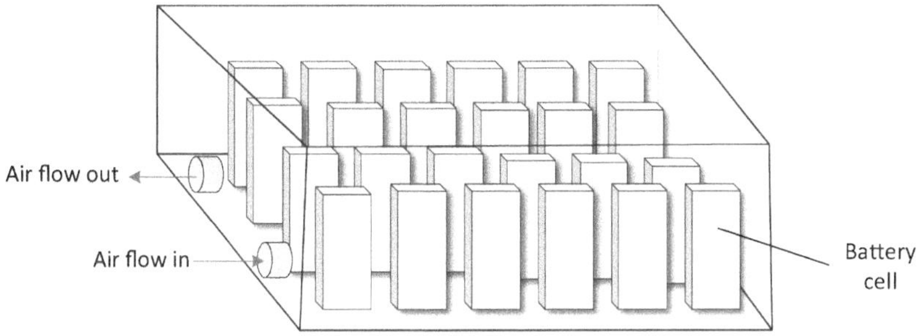

(a) Horizontal layout of
battery pack

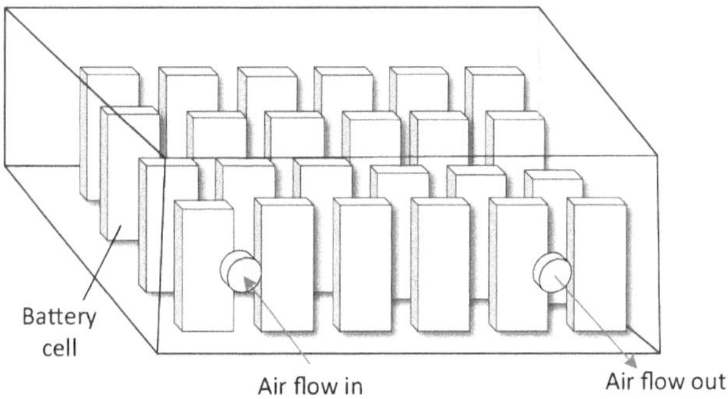

(b) Longitudinal layout
of battery pack

Fig. 5.3 Horizontal and longitudinal layouts of battery packs in which the inlet and outlet ports for the air coolant are exposed to the respectively larger and thinner surface areas of the cells.

In the parallel air-flow configuration of an air-based battery thermal management system, increasing the inlet air flow can reduce the battery pack temperature, but alone it has no significant effect on temperature uniformity of the battery pack. The performance of a parallel air flow configuration can be improved using a secondary vent (see Fig. 5.5). The location and size of the secondary vent strongly affects the maximum temperature of the cell and the temperature uniformity within the battery pack. It is recommended to place the secondary vent near the cell with the maximum temperature (Hong et al. 2018). Increasing the vent size decreases the cell maximum temperature (Hong et al. 2018).

An axial air flow direction can be used in a battery thermal management system when the battery pack contains cylindrical batteries. When a forward axial air flow direction is used in such a battery thermal management system

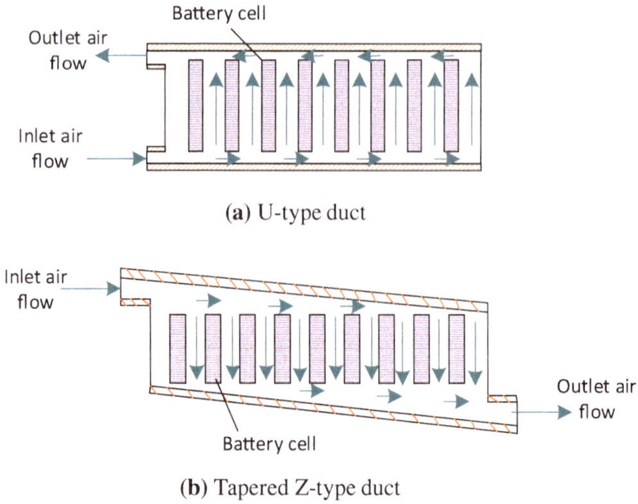

Fig. 5.4 U-type and tapered Z-type air ducts for battery pack in an air-based battery thermal management system.

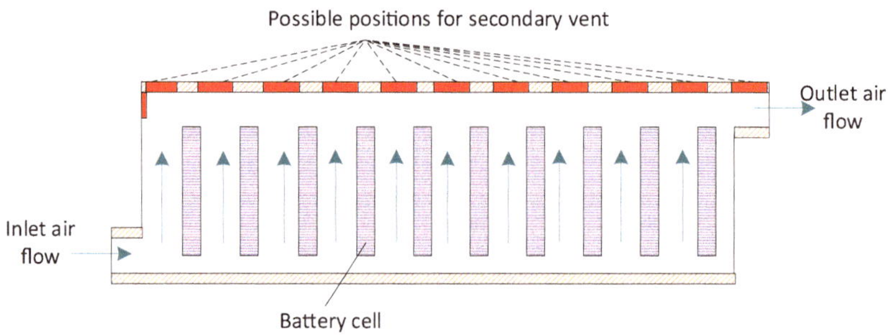

Fig. 5.5 Z-type air vent for a battery pack with possible positions for secondary vent.

(Fig. 5.6), the cylindrical batteries are arranged in an axial orientation and air flows through them via channels in the middle of the battery pack. Increasing the distances between the battery cells slightly raises the average temperature of the battery pack, which is not desired, but it helps in improving the temperature uniformity in the battery pack (i.e., it decreases the temperature differences between the cells) (Yang et al., 2016a, b). In addition, the power required for providing cooling for the battery pack increases as the radial interval of battery pack increases. Moreover, the cell temperature decreases with an increase in the air flow rate when using the forward axial air flow configuration.

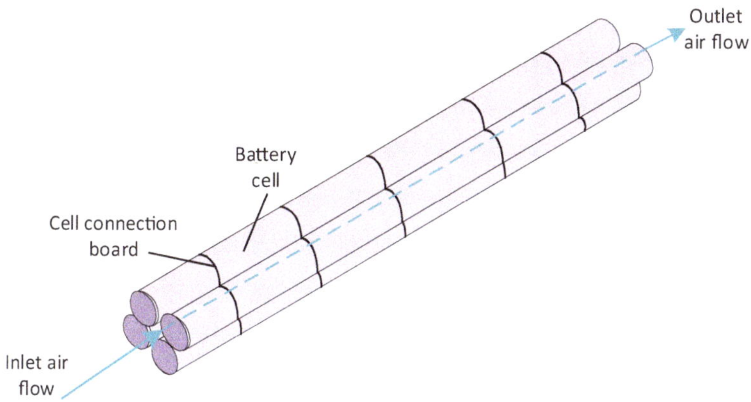

Fig. 5.6 Axial air flow cooling for a battery pack containing cylindrical batteries.

Numerous design configurations for a forced air convection battery thermal management system were investigated by Wang et al. (2014). Various cross section areas and shapes (square, rectangular, hexagonal, and round) were considered for the battery pack, as seen in Fig. 5.7. There, the arrangement of battery cells is shown from the top view of the battery pack. A fan is used to provide forced air convection through the cylindrical batteries, and the air flow inlets are shown in Fig. 5.7. Forced air flows axially through the square pack design and, for the rectangular design, air enters from one side and leave the channel from the other side. Depending on the arrangement employed, some cells are cooled well and others are cooled less adequately or inadequately. In the rectangular arrangement of battery cells, the battery cells in the middle are not effectively cooled, while in the square design, the cells close to the walls of the battery pack are not effectively cooled. In addition to battery pack design, other design parameters like distance between cells and temperature and speed of the inlet air flow significantly affect the cooling of the cells in the pack.

The outlet air from a battery system can be cooled in an air cooling system and then returned the battery system or discharged to the environment, i.e., not recycled. An air cooling system can be designed to remove heat an air space, like a car cabin. In the case in which cooling air is not recycled (see Fig. 5.8), the air is discharged to the environment by one or more fans after cooling the batteries. Such air-cooled systems are relatively small as no space is required to recycle the air. A commonly used cabin air cooling system is the vapor compression refrigeration cycle, which is comprised of an evaporator, a compressor, a condenser and an

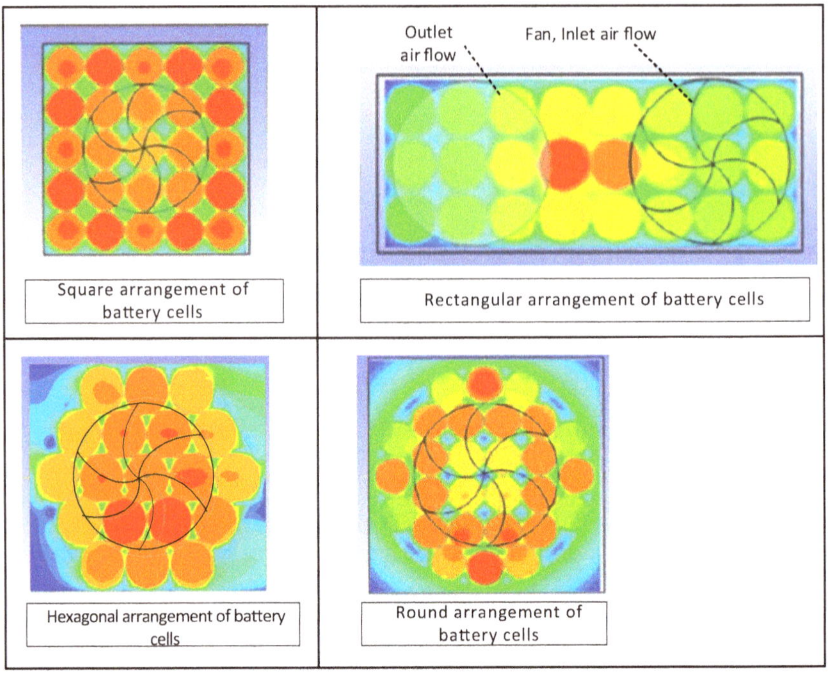

Fig. 5.7 Various arrangements for cylindrical batteries for use with forced convection air battery thermal management systems. Cross section areas of battery packs are shown. The more red the batteries, the higher is their temperature. *(Adapted from Wang, T., Tseng, K.J., Zhao, J., Wei, Z., 2014. Thermal investigation of lithium-ion battery module with different cell arrangement structures and forced air-cooling strategies. Appl. Energy 134, 229–238.)*

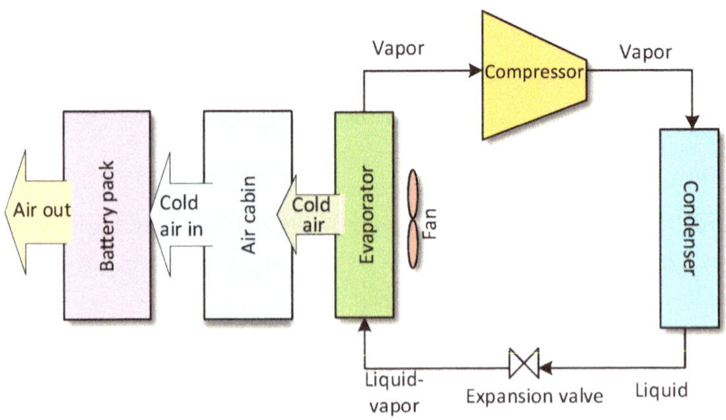

Fig. 5.8 Cabin air cooling system using vapor compression refrigeration cycle to cool a battery pack.

expansion valve. The relatively low temperature air is blown through ducts and manifolds to the battery system by a fan. In the case in which air is recycled, heat is transferred from the relatively high-temperature air leaving from the battery to the evaporator, so the air can return to the battery system to absorb heat. In some applications, like electric and hybrid vehicles, air flow over the battery system for cooling can be induced through movement of the vehicle. Although this type of air-based cooling system uses no power to move the air, it is constrained to moving devices with battery pack like electric and hybrid vehicles.

Air-based battery thermal management systems are not without their challenges. Although air-based battery thermal management systems are small in size and consume relatively little power to reject heat from a battery system, they are not effective at providing an even temperature distribution among the cells in the battery pack, in part because of the low heat capacity and low thermal conductivity of air. Also, air cooling systems can provide cooling only through sensible heat removal, which is often not sufficient for high-powered battery applications. Trying to increase the air flow rate to provide a high cooling effect for high-powered battery applications where significant heat is generated can be problematic, as doing so usually leads to large ducts and manifolds and increase and high power consumption by fans. Furthermore, air-based thermal management systems can be noisy as they usually work with fans and blowers. As a consequence, other types of battery thermal management systems that are not air-based systems compared have been developed.

5.5 Liquid-based battery thermal management systems

The weaknesses of air-based battery thermal management systems, especially for high heat loads, have led to the development of liquid-based battery thermal management systems. These are particularly well suited for relatively high-powered battery applications like electric and hybrid electric vehicles. Liquid-based cooling systems are the most commonly used battery thermal management systems for electric and hybrid electric vehicles.

Liquid-based battery thermal management systems can be generally classified based on whether the surface of the battery is in *direct* or *indirect contact* with the heat transfer fluid. In the direct contact systems, a dielectric heat transfer fluid (like mineral oil), chosen to eliminate the risk of short circuiting, is typically to remove heat from the battery. This type of thermal management system has the advantage of higher cooling rates and greater compactness compared to the indirect liquid-based battery thermal management systems. Some direct contact liquid-based cooling systems using boiling phase change phenomenon are described in Section 5.5. However, the usage of a direct contact-based heat transfer fluid is not practical for some battery applications. Instead, indirect contact liquid-based cooling systems can be straightforwardly employed to dissipate heat from a battery system. Higher heat transfer fluid flow rates can be used in indirect contact-based systems as the working fluid typically has a lower viscosity compared to that of the heat transfer fluid in the direct contact-based thermal management systems.

Generally, a heat transfer fluid with enhanced thermal conductivity and heat capacity is preferable for liquid-based battery thermal management systems. Various heat transfer fluids can be used in indirect contact-based cooling systems, including water, ethylene glycol-water solution, deionized water with sodium polyacrylate, mineral oil, Al_2O_3-water nanofluid, liquid metal and the refrigerant R-134a. Although liquid water is commonly used as the heat transfer fluid for the liquid-based cooling systems, it has several challenges, including low thermal conductivity and freezing at subzero climatic conditions. These drawbacks of water can be resolved by adding antifreeze (e.g., ethylene glycol) and materials like aluminum oxide (Al_2O_3) or nanomaterials to enhance the thermal performance of water.

With indirect contact-based thermal management systems, the heat transfer fluid is either pumped through channels built in a *metal plate* (*cold plate*) or pumped through *discrete tubes*. The cold plate can be fabricated in a flat shape (for prismatic batteries) or in cylindrical shape (for cylindrical batteries), and it can also serve as a structural support for the batteries. The cold plate can be placed into the battery, between the cells (sandwich configuration) or to the sides of battery module. Fig. 5.9 shows various placements for flat cold plates in the prismatic batteries. For internal placement of the cold

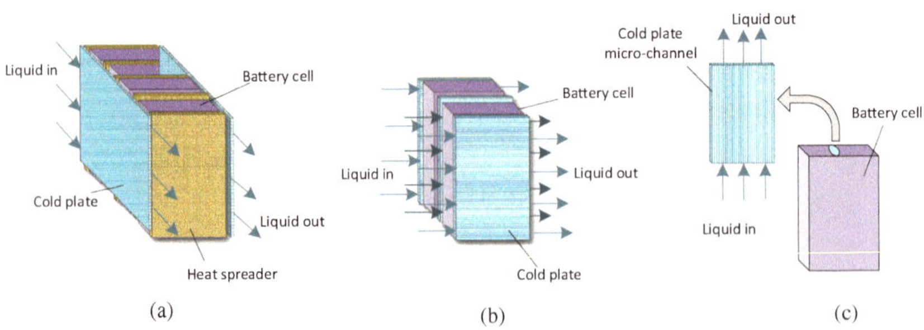

Fig. 5.9 Cold plate placement options for indirect contact-based thermal management systems a cold plate: (A) on module sides, (B) between battery cells, (C) in battery cell.

plate, the size of channels should be sufficiently small to permit them to be embedded in the components of the battery. Also, the cold plate and its channels should be composed of chemically inert materials to prevent possible chemical reactions with the components of the battery cell. When the cold plates are placed to the sides of the battery module, two heat spreaders (which transfer heat from the battery to a colder heat sink) can be used to effectively transfer heat from the module to the cold plates. For the sandwich configuration of cold plates, where they are placed between battery cells, the thickness of cold plates should be small enough to facilitate better configurations of these plates between the battery cells. The flow patterns and number of channels of a cold plate can be designed based on maximizing heat transfer effectiveness and minimizing the pressure drop within the channels.

Fig. 5.10 shows typical configurations of channel designs for an indirect contact-based thermal management system using a cold plate. The main configurations are parallel and serpentine. In the parallel configuration of channels in a cold plate, an increase in the number of channels results in better cooling performance. However, there is an optimum number of channels, above which no apparent improvement is obtained in cooling performance. In addition, with a simultaneous increase in the width of channels and cooling medium mass flow rate, the battery pack maximum temperature, as well as the temperature differences between the cells, are reduced. The pressure drop is much greater in the serpentine configuration of the cold plate channel than in the parallel configuration. In the serpentine design of channels in a cold plate, increasing the radius of the channel bend and the width enhances the cooling process and reduces the pressure loss. In addition, the temperature uniformity of battery pack can be improved by

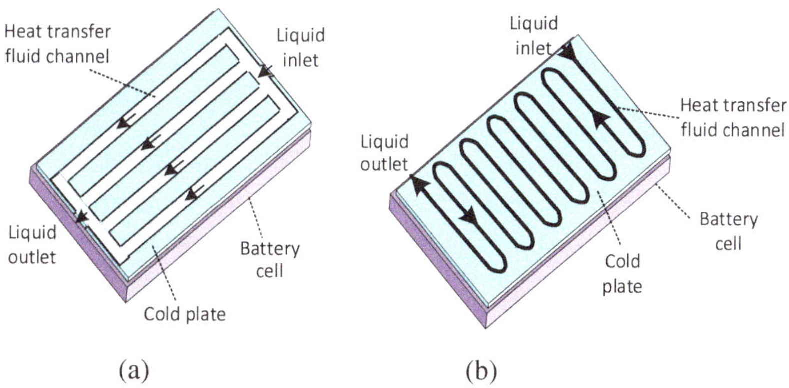

Fig. 5.10 (A) Parallel and (B) serpentine channel configurations for a cold plate.

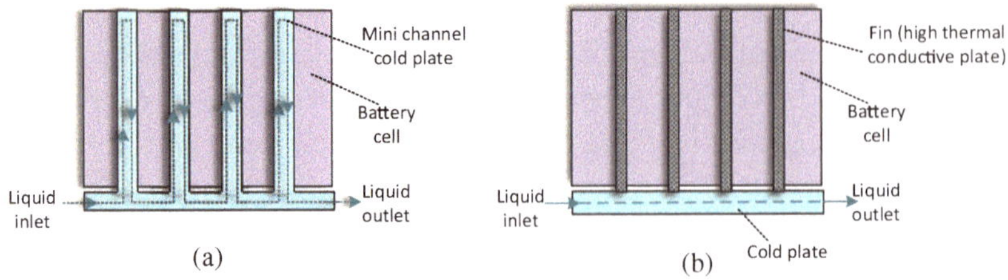

Fig. 5.11 Use of (A) mini-channel cold plate and (B) high thermal conductive plates (fins) for cooling battery cells.

widening the inlet channel toward the outlet. This provides a balance between the heat transfer area, coolant velocity and solid-fluid temperature gradient, which leads to an equal heat transfer for all plate areas. Also, in the parallel configuration, divergent-shaped channels are an effective design property in which the maximum temperature of the battery pack and the pressure drop within the cold plate is minimized.

The mini channel cold plate cooling method (see Fig. 5.11A) can also be used to remove heat from battery cells. In this method, a set of mini channel cold plates with high thermal conductivities are placed between the cells and directly attached to them. This method is capable of providing a uniform temperature distribution throughout the battery pack. Since narrow and long channels are used for the flow of liquid coolant with this method, the overall pressure drop is relatively high. Consequently, electric pumps are required to pump higher flow rates of liquid coolant at higher static pressures, and this results in higher electrical power use for this method. Thus, optimization of mini channel cold plate cooling systems is necessary to determine the optimum number and geometry of channels, liquid flow rate and direction, and operating conditions for the cold plate.

Another type of indirect liquid-based battery thermal management system involves the use of fins between battery cells. For this fin-based cooling method (see Fig. 5.11B), fins with high thermal conductivity are attached between the battery cells. The fins transfer the heat of each cell to the liquid coolant flowing in a cold plate. The fluid is then discharged to the outside. This method is able to provide a uniform temperature at the cell level only. In fact, the temperature difference between the cell and the fin gradually decreases, resulting in a reduction in cooling performance. In addition, since the liquid flows forward (from the first cell to the last), its temperature increases as it reaches to the last cell. Therefore, a uniform temperature distribution cannot be attained throughout the battery pack with this method of cooling.

The design and configuration of a liquid-based cooling plate system that absorbs heat from the battery pack is described in the earlier part of this section. To extend that treatment, a secondary cooling cycle is now described that can be used to cool the liquid coolant exiting the battery pack. Fig. 5.12A shows a typical secondary cooling system. It includes two loops: a refrigerant loop and a liquid coolant loop. A two-way valve is used to adjust the fluid flows to match the cooling loads. The following settings can be used for the two-way valve:

(i) For relatively high cooling loads, the valve is set to allow the liquid (i.e., cooling medium) to enter the chiller and be cooled by the compression refrigeration cycle.

(ii) For relatively low cooling loads and when the environment temperature is lower than the liquid temperature, the two-way valve is set to allow the liquid to pass through the radiator, where ambient air is used to cool the liquid.

In the case of heating a battery at very low temperature operating conditions (e.g., during winter season), Fig. 5.12A can be modified to add a heat pump function to the system as illustrated in Fig. 5.12B. Then, the battery can be heated and also cooled by changing the direction of working fluids using the three-way valve. The secondary cooling type and design for liquid-based battery thermal management systems are selected, noting that it is important to maintain the battery temperature at the desired operating temperatures during various power loads and environmental conditions.

With proper design of liquid-based battery thermal management systems, the desired battery operating temperature can be achieved even under extreme operating and/or environmental conditions, such as rapid discharging or charging at very hot or cold conditions. Liquid-based battery thermal management systems are compact in size but they are complex. They have additional components (e.g., coolant piping), which often increase the weight and risk of leakage and consequently lead to more extensive maintenance requirements.

Several important advances in liquid-based battery thermal management systems have been reported recently. These typically aim to improving cooling performance for various operating conditions. Basu et al. (2016) presented a new liquid-based cooling system based on the concept of a heat pipe in which an aluminum conducting element is used for a set of lithium ion battery cells. The use of the conducting element reduces the contact

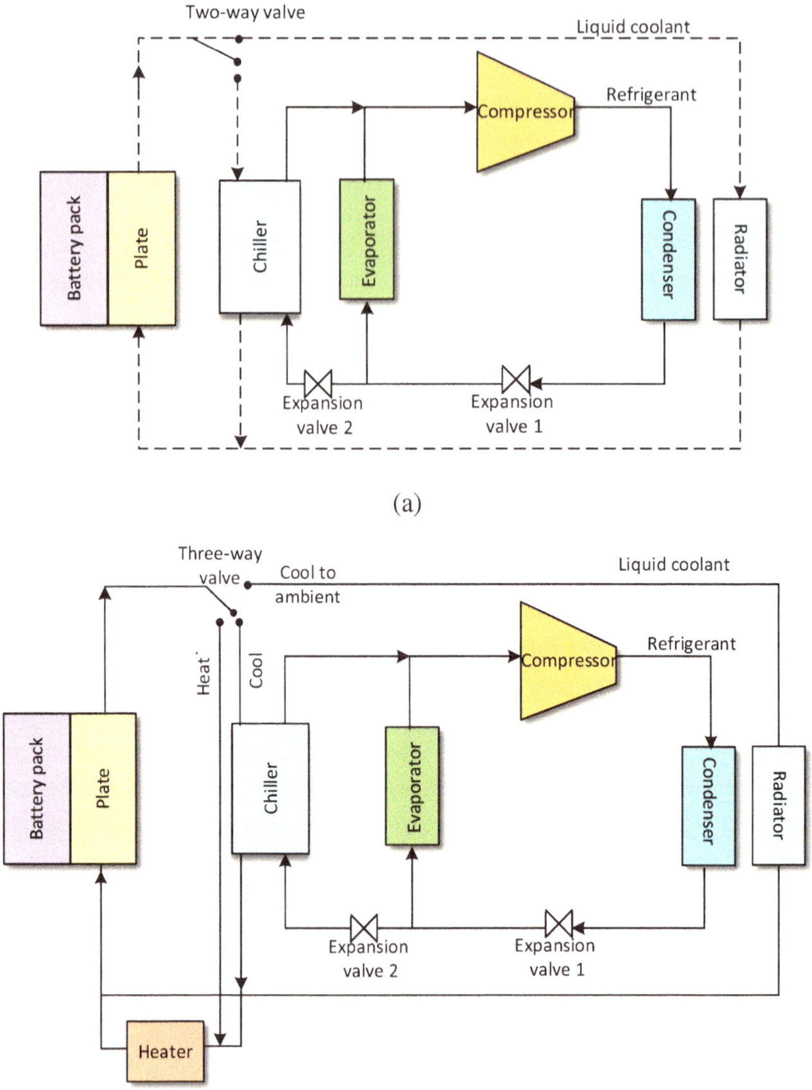

(b)

Fig. 5.12 (A) Secondary cooling system including refrigerant loop and liquid coolant loop for cooling a battery pack. (B) Secondary cooling (including refrigerant and liquid coolant loops) and heating systems for cooling and heating a battery pack.

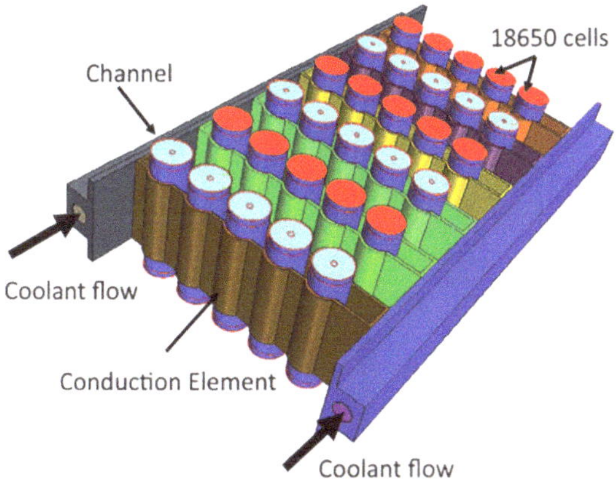

Fig. 5.13 Geometry of battery pack combined with liquid and conducting element battery thermal management system. *(From Basu, S., Hariharan, K.S., Kolake, S.M., Song, T., Sohn, D.K., Yeo, T., 2016. Coupled electrochemical thermal modelling of a novel Li-ion battery pack thermal management system. Appl. Energy 181, 1–13.)*

resistance between the coolant and the battery cell and also serves as a separator between the coolant and the cell, preventing direct contact between the two in the case of leakage or other system failures. Fig. 5.13 illustrates this battery pack, integrated with a liquid coolant and conducting element thermal management system. This battery thermal management system was found to be capable of cooling the battery pack even at low flow rates of the liquid cooling medium. In addition, the system can cool the battery pack effectively for undesirable operating conditions (e.g., low flow rate of coolant and high discharge rate), making it a promising candidate for electric vehicles.

The use of liquid metal as a coolant instead of water in liquid-based battery thermal management systems was proposed by Yang et al. (2016a, b). They showed that, at similar flow conditions, the use of liquid metal in a battery thermal management can provide a more uniform temperature distribution across the battery pack while consuming less power compared to the use of water in that battery thermal management system. The liquid metal is capable of removing heat from the battery pack for relatively high-powered applications, hot weather conditions and some cell failure conditions. Although a liquid metal is a potentially promising coolant for liquid-based battery thermal management systems, the high mass of this coolant (due to the high density of liquid metal) makes such a system relatively heavier compared to conventional water-based thermal management

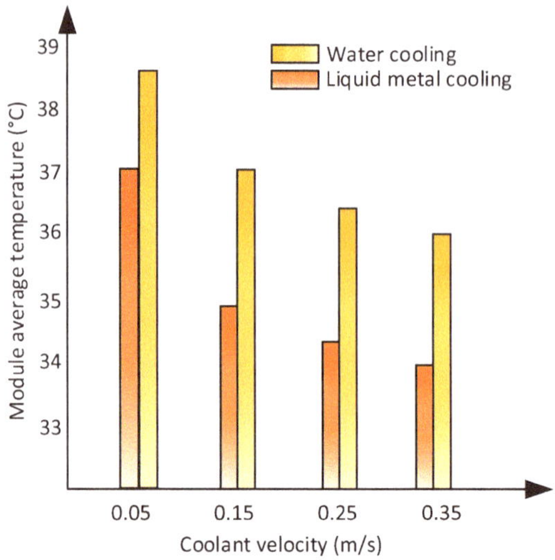

Fig. 5.14 Average temperature of four lithium ion battery cells for various coolant velocities and types in a liquid-based battery thermal management system. *(Data extracted from Yang, X.H., Tan, S.C., Liu, J., 2016a. Thermal management of Li-ion battery with liquid metal. Energy Convers. Manag. 117, 577–585. Yang, T., Yang, N., Zhang, X., Li, G., 2016b. Investigation of the thermal performance of axial-flow air cooling for the lithium-ion battery pack. Int. J. Therm. Sci. 108, 132–144.)*

systems. Fig. 5.14 compares, for a specific liquid-based battery thermal management system, the temperature of liquid metal and water at several coolant velocities. The cooling system using liquid metal is seen to exhibit better thermal performance than the water-based cooling system.

Thermal management systems should not only provide the desired operating temperature for the battery system, but they also should be designed to control the potential for thermal runaway in the battery system. The phenomenon of thermal runaway occurs when the battery temperature exceeds from the critical temperature (i.e., maximum allowable operating temperature). When the battery thermal management systems do not effectively dissipate enough heat from the battery system, the temperature of a battery rises and approaches the critical temperature. In addition, internal short circuiting in batteries, which can result from improper battery fabrication and/or vehicle collisions, can lead to thermal runaway. Appropriately mitigating the potential for thermal runaway is important because the phenomenon can ignite a fire and lead to explosion of the battery. Numerous efforts for controlling thermal runaway in batteries have been established and new ones are being considered via the development of innovative designs and configurations for the battery thermal management systems. For example, Xu et al. (2017) proposed a mini channel liquid-based thermal management system for a battery module. The system was analyzed under internal short circuit conditions of the battery. Despite the fact that the proposed mini channel cooling system was unable to halt thermal runaway caused by an internal short circuit, it was able to prevent the dispersion of thermal runaway among battery cells.

5.6 Phase change material (PCM)-based battery thermal management systems

A PCM uses the latent heat of changing from one phase to another to absorb or release thermal energy, which can be used for heating or cooling. As stated earlier, liquid-based battery thermal management systems can effectively maintain the temperature of a battery within the desired range, but this often requires the use of a notable amount of energy. In addition, the incorporation of piping, pumps and channels in the cooling loops of liquid-based thermal management systems often make these systems complicated. A PCM-based thermal management system, however, can be considered a passive system that can store and/or release large amounts of thermal energy with no additional energy consumption. Also, a PCM-based system can provide greater temperature uniformity since the latent heat is used to dissipate heat from the battery rather than the sensible heat as is used in air/liquid-based battery thermal management systems. During the sensible heat exchange process, the temperature of heat transfer medium changes continuously, while in latent heat process, the temperature remains constant during the heat absorption from each single cell as well as battery pack, which leads to high temperature uniformity. Fig. 5.15 shows temperature characteristics in a battery thermal management system with/without PCM incorporated into the system. It is seen that, in the PCM-based battery thermal management system, the temperature of the system remains constant at the melting temperature of the PCM, and then increases after the completion of the phase change process. This provides a higher potential for the thermal management system to absorb heat from the battery pack compared to the air/liquid-based thermal management system, in which the temperature of the system increases continuously during the sensible heat exchange between the air/liquid and battery pack.

A simplified diagram of a PCM-based battery thermal management system is shown in Fig. 5.16. In this system, the battery cells are attached to the PCM that allows the cells to be attached. Two plates are attached at the right

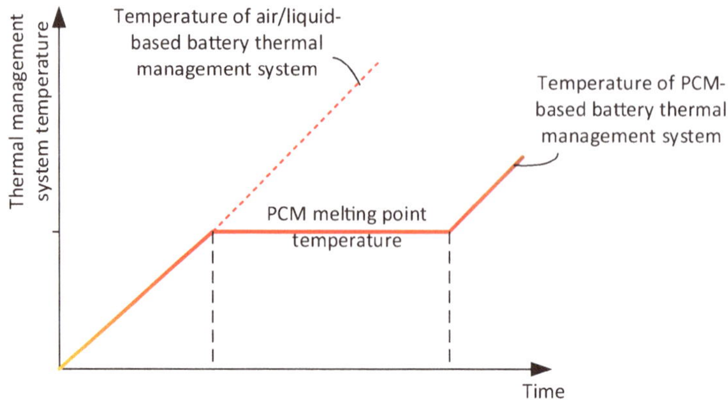

Fig. 5.15 Temperature profiles in PCM-based and air/liquid-based battery thermal management systems.

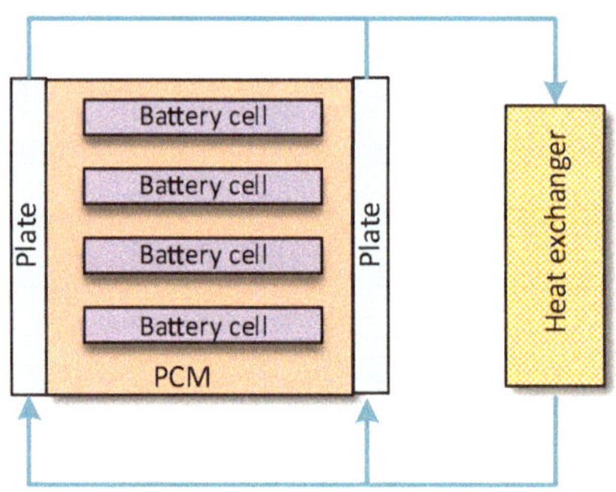

Fig. 5.16 Simplified PCM-based battery thermal management system.

and left sides (or the top and bottom) of the PCM to remove the absorbed heat earlier by the PCM. During the charge and discharge of the battery system, the generated heat in each battery cell is transferred to the PCM. The temperature of the PCM increases at first as it absorbs heat from the battery cells until it reaches the PCM melting point. It then absorbs a significant amount of heat at constant temperature as the phase change proceeds (e.g., the solid PCM is converted to a liquid). Note that a PCM-based thermal management system cannot operate effectively in continuous operation of the PCM during charge or discharge, or at high environmental temperatures, because the PCM eventually becomes completely melted. Increasing the mass of PCM for storage of more heat in the PCM increases notably the mass of the PCM-based thermal management system and consequently increases the powered consumption in some battery applications like electric vehicles. So, a secondary cooling system is required to transfer the heat of the PCM to the outside and to solidify the melted PCM. The secondary cooling system can restore the heat storage capacity of the PCM through forced or natural air convection cooling or liquid cooling or some other means. A number of studies have focused on the secondary cooling system for PCM-based battery thermal management systems (Sabbah et al., 2008; Ling et al., 2015; Wu et al., 2017a, b; Zhao et al., 2017; Hémery et al., 2014; Duan and Naterer, 2010; Wu et al., 2016; Javani et al., 2014).

In addition to using a proper mass of PCM in a PCM-based battery thermal management system, there are some other PCM requirements or preferences. For instance, the PCM should be nontoxic, chemically stable and resistant to subcooling effect in the freezing process (Jaguemont et al., 2018). Furthermore, the selection of a suitable PCM for a PCM-based battery thermal management system is important for achieving efficient cooling performance. Ideally, it is advantageous for the PCM to have a high heat capacity, latent heat, melting temperature and thermal conductivity. In reality, however, the low thermal conductivities of PCMs prevent fast responses from PCM-based thermal management system to high cooling loads. For example, paraffin is often considered a suitable PCM because of its low cost, high latent heat and suitable phase change temperature for many applications; however, it has a low thermal conductivity like most PCMs.

Various methods can be utilized to increase the thermal conductivity of a PCM. One of the methods to increase the thermal conductivity of a PCM is adding to it a material with a high thermal conductivity. A number of tests have been reported of the use in PCM-based battery thermal management systems various of materials with high thermal conductivities. For example, Mehrali et al. (2016) examined the use of carbon fibers, Shirazi et al. (2016) carbon nanotubes, Goli et al. (2014) and Mehrali et al. (2016) graphene, Huang et al. (2015) and Li et al. (2014) metallic foam/mesh, and Malik et al. (2016) metal particles. Other methods to improve the thermal conductivity of PCMs include the use of porous material (e.g., graphite matrix) and the use of metal fins, as described by Pan et al. (2016).

The thermal characteristics for several composite paraffin-based PCMs are presented in Table 5.1. Specifically, the melting point, latent heat and thermal conductivity for selected composites of paraffin are presented. It is seen that adding ammonium polyphosphate (APM), high density polyethylene (HDPL), and primary sclerosing cholangitis (PSC) to the pure paraffin increases the melting temperature and thermal conductivity of the paraffin composites, while reducing the latent heats of the composites.

Table 5.1 Melting point, latent heat and thermal conductivity for selected composites of paraffin.

PCM (additive mass fraction)	Melting temperature (K)	Latent heat (kJ/kg)	Thermal conductivity (W/m^2 K)
Paraffin[1]	319–321	173.4	0.12–0.21
Paraffin (60%)/HDPE (20%)/APM (20%)[2]	347.7	50.58	0.29
Paraffin (60%)/HDPE (40%)[3]	341.3	51.59	0.28
Paraffin (60%)/HDPE (15%)/APM (20%)/EG (5%)[2]	346.6	50.58	0.85
Paraffin (90%)/EG (10%)[2]	313.2	178.3	0.82
Paraffin (75%)/PSC (25%)[4]	329.3	165.16	0.387

Abbreviations: APM, ammonium polyphosphate; EG, expanded graphite; HDPL, high density polyethylene; PSC, primary sclerosing cholangitis.
Data sources: [1]Kandasamy et al. (2007); [2]Al-Zareer et al. (2018a); [3]Zhang et al. (2010); [4]Zhou et al., (2009).

Research is ongoing to enhance further the thermal conductivity of many PCMs and to resolve such other challenges as low heat transfer area between the PCM and the outside, and leakage in PCM-based battery thermal management systems. To address the leakage issue, Wu et al. (2019) indicate that a new composite PCM that is stable and has thermally induced flexibility can be used in PCM-based thermal management systems. In addition, the integration of a PCM with air forced convection and the utilization of fins can increase both the heat transfer coefficient and surface heat transfer area.

Another active area of research on PCM-based battery thermal management systems is on improving performance by integrating various cooling systems with battery thermal management systems. For example, a heat pipe for heat transport is integrated with a PCM to add the characteristics of high thermal conductivity to the battery thermal management system (Wu et al., 2017b). The use of the heat pipe enhances both the cooling process for the battery pack and the charging and discharging rates of the PCM. The operating principle of the heat pipe is described in Section 5.7.1. A PCM-based battery thermal management system integrated with a heat pipe is illustrated in Fig. 5.17. It is seen that the batteries are surrounded by PCM composites that absorb and store the heat of the batteries though sensible and/or latent heat processes. The evaporating part of the heat pipe is connected to the PCM and its condensing section extends outside of the battery pack. In the condenser section of the heat pipe, a number of fins are used and they are in in thermal contact with an air flow channel to condense the vapor

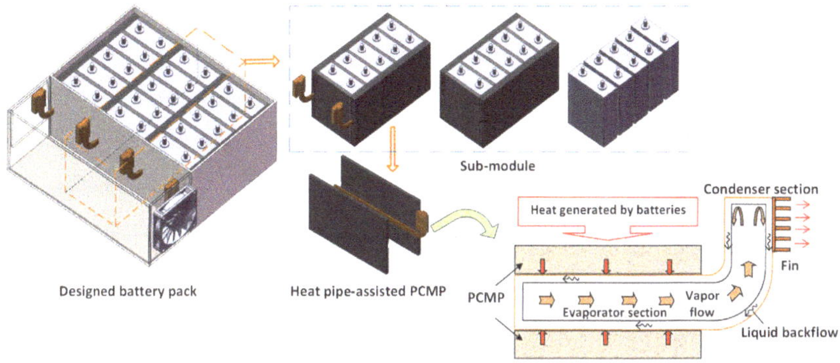

Fig. 5.17 Design of the battery pack proposed by Wu et al. (2017b) in which heat pipes are used in the PCM-based battery thermal management system to improve the cooling process of the batteries. *(Modified from Wu, W., Yang, X., Zhang, G., Chen, K., Wang, S., 2017b. Experimental investigation on the thermal performance of heat pipe-assisted phase change material based battery thermal management system. Energy Convers. Manag. 138, 486–492.)*

flow. Increasing the velocity of air can lead to more effective forced convection, which can control the battery temperature at relatively high discharge rates. In fact, the use of a heat pipe improves the PCM heat absorption and dissipation rates and subsequently increases the thermal performance of heat pipe-assisted PCM-based battery thermal management systems.

Furthermore, a tube-shell heat exchanger can be integrated into a PCM-based battery thermal management system (Jiang et al., 2017), as seen in Fig. 5.18. The use of the tube-shell heat exchanger cools the PCM, allowing it to provide a uniform temperature throughout the battery pack, with a maximum temperature difference of 1–2°C. It can be seen in Fig. 5.18 that the heat of cylindrical batteries is absorbed by the PCM composite and then transferred to the forced air circulating through the heat exchanger. In addition, the use of baffles can change the air flow directions, increasing interactions between the batteries and air. This can ultimately enhance the heat transfer performance of the integrated shell and tube heat exchanger with PCM-based battery cooling system.

5.7 Liquid-vapor phase change-based battery thermal management systems

5.7.1 Heat pipe-based battery thermal management systems

The previous section describes the PCM-based battery thermal management system and points out that that type of system does not consume additional

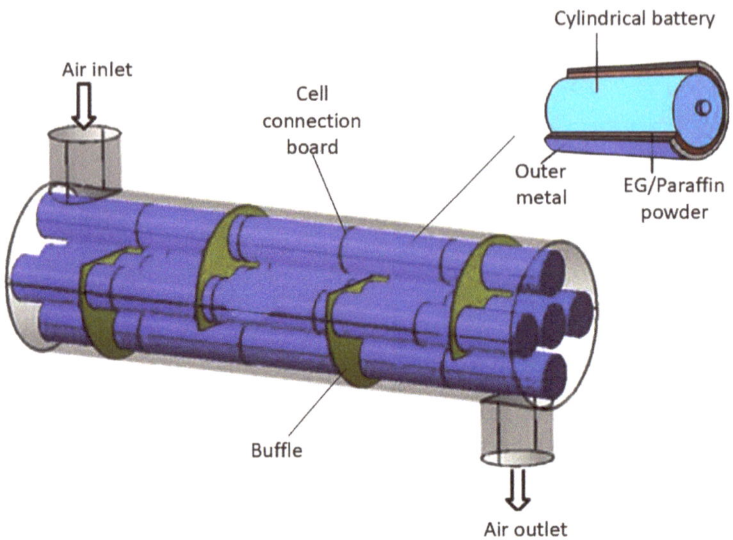

Fig. 5.18 Shell and tube heat exchanger integrated with PCM for cooling cylindrical batteries. *(Modified from Jiang, G., Huang, J., Liu, M., Cao, M., 2017. Experiment and simulation of thermal management for a tube-shell Li-ion battery pack with composite phase change material. Appl. Therm. Eng. 120, 1–9.)*

power, but it cannot operate effectively in lowering the temperature difference within a battery pack continuously. In addition, the volume variation that occurs during the phase change process imposes additional limitations for use of such systems. As an alternative, a heat pipe is a heat transfer device that can operate without an external pump and is capable of removing heat at significant distances with high speed from various types of devices, including batteries. Heat pipes have been widely used recently in various thermal management applications due to their ability for effective heat transfer, compact structure, flexible geometry and long lifetimes. However, their capacities and surface areas need to increase for use in high-power battery operations like electric vehicles.

A heat pipe generally involves three processes, each of which occurs in a separate section: evaporation section, adiabatic section and condensation section (see Fig. 5.19). A working fluid flows through the heat pipe. The working fluid absorbs heat in the evaporation section, which is in contact with the external heat source (e.g., battery). Then the vaporized working fluid is transferred to the condensation section, after passing through the adiabatic section, due to internal pressure difference through the container. In the condensation section, the working fluid is condensed by an external

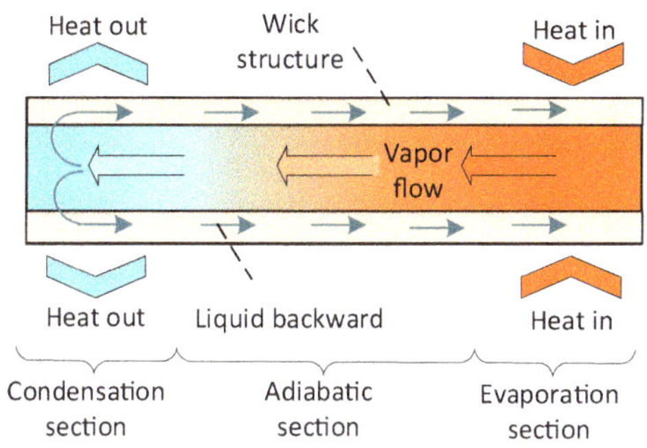

Fig. 5.19 Working principle of heat pipe.

cooling system and then returns to the evaporation section without external power consumption through the capillary force of the wick.

A heat pipe-based battery thermal management system is shown in Fig. 5.20 along with the secondary cooling system used to remove heat from the condensation section of the heat pipe. Several heat pipes are used. The evaporation sections are in contact with the batteries where the heat pipe working fluid turns is vaporized as it absorbs battery heat. Then vapor condenses in the condensation section where a secondary coolant is used to remove heat from the heat pipe working fluid. Since the surface contact area between a given heat pipe and the battery is small, heat pipes can be coupled with a cold plate which has high thermal conductivity to facilitate the transfer of heat from battery and its dissipation via an external air- or liquid-based cooling system.

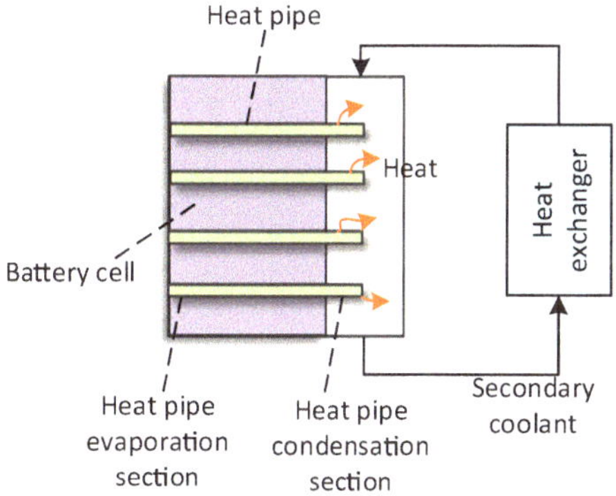

Fig. 5.20 Heat pipe-based battery thermal management with its secondary cooling system.

The thermal performance of heat pipe-based battery thermal management systems can be enhanced by proper selection of the working (heat transfer) fluid, appropriate design and configuration of the heat pipe and employment of a suitable cooling method for the condensation section. The shape and size of the heat pipe should be designed so as to yield a high degree of contact between the evaporation section of the heat pipe and the battery cells. For instance, an evaporator with a flat shape can provide higher heat transfer area compared to one with a tubular shape. A metal plate can be also connected with the heat pipe to effectively distribute heat and to maintain a more constant temperature. An integrated heat pipe and metal plate can be placed on the battery module or between the battery cells. Furthermore, an effective condensation process is required in the condenser section for successful implementation of a heat pipe. The condenser section can be extended and equipped with fins to increase the heat transfer area (see Fig. 5.21). A fan is typically used to create a forced air flow to dissipate heat from the fins, better facilitating condensation of the working fluid in the condensation section of the heat pipe. Other approaches have been developed or proposed to remove heat from the fins. These include the use of natural air convection, wet cooling and thermal baths for cooling, as well as the use of alternative cooling mediums like water or glycol-water mixture.

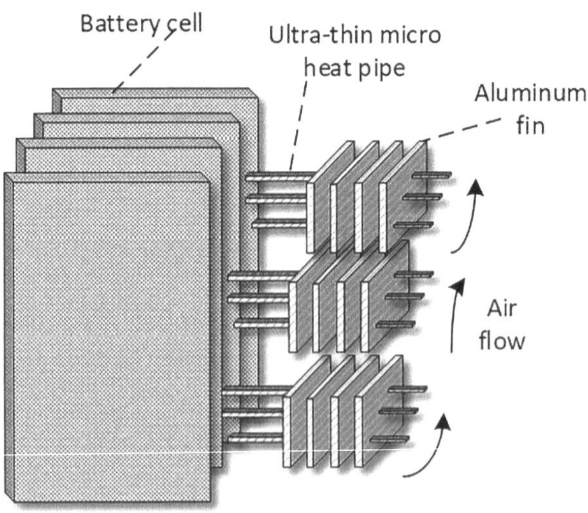

Fig. 5.21 Use of fins with a heat pipe to increase the convection surface area with air and thereby enhance battery cooling.

The performance of a heat pipe strongly depends on its installation angle. For most heat pipe applications, the vertical placement of heat pipes, with the evaporator and condensation sections respectively at the bottom and top of the heat pipe, results in the desired heat pipe performance. With this orientation, the condensed working fluid flows down to the evaporator section by gravity.

Numerous studies for performance improvement of heat pipe-based battery thermal management systems have reported. Most are applicable to prismatic battery cells. Further developments and research are needed on heat pipe-based systems for cylindrical batteries.

5.7.2 Evaporating pool boiling-based battery thermal management systems

Evaporating pool boiling-based systems for battery thermal management are relatively new. These systems are promising, as they are capable of supplying higher cooling loads for batteries than conventional cooling systems without the need for an external power supply. In evaporating pool boiling-based battery thermal management systems, the batteries are directly submerged in a stationary liquid heat transfer fluid. This fluid absorbs battery heat and its temperature rises to the boiling point. Thereafter, the heat transfer fluid absorbs heat at constant temperature as vaporization occurs. The heat transfer fluid is selected to have a boiling temperature near to the desired operating temperature of the battery.

Various heat transfer fluids have been considered and compared for the thermal management of lithium ion batteries in a hybrid electric vehicle, including propane (Al-Zareer et al., 2017b), ammonia (Al-Zareer et al., 2017a), and refrigerant R134a (Al-Zareer et al., 2018a). In those studies, the ammonia and propane heat transfer fluids are not recirculated through the battery pack, as they are consumed as fuels for the hybrid electric vehicle. For the system using refrigerant R-134a as the heat transfer fluid, the R-134a vapor generated during boiling of the refrigerant is condensed by the vehicle cabin air cooling system. Fig. 5.22 shows a schematic of a proposed evaporating pool boiling-based battery thermal management system using ammonia as the heat transfer fluid. The heat transfer fluid is first used for cooling batteries then fed to the vehicle engine. The authors claim a high heat transfer coefficient, a large thermal capacity (due to using the latent heat of the heat transfer fluid) and a high thermal conductivity for such evaporating pool boiling-based battery thermal management systems. These systems can maintain the battery temperature within the desired range.

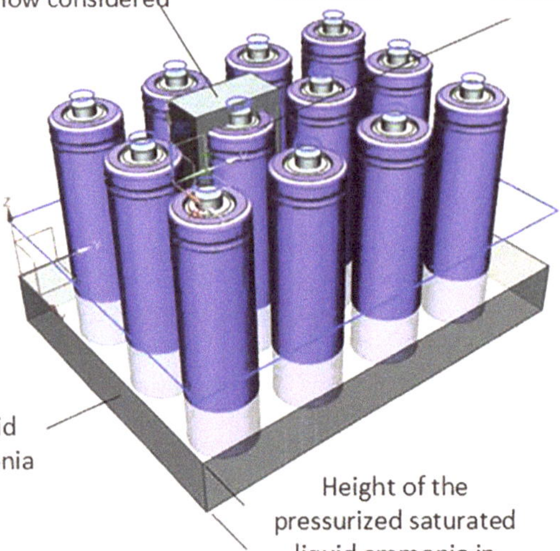

Fig. 5.22 Evaporating pool boiling-based battery thermal management system using stationary ammonia as the coolant. *(From Al-Zareer, M., Dincer, I., Rosen, M.A., 2017a. Electrochemical modeling and performance evaluation of a new ammonia-based battery thermal management system for electric and hybrid electric vehicles. Electrochim. Acta 247, 171–182.)*

An experimental pool boiling-based battery thermal management system was investigated by Van Gils et al. (2014). In that system, a cylindrical battery is completely submerged in Novec7000 heat transfer liquid, which has a boiling temperature of 34°C at 1 atm. This heat transfer fluid is a dielectric, so it does not cause short circuiting while removing heat from the battery. Fig. 5.23 illustrates the experimental setup of Van Gils et al. (2014). During charge and discharge, when heat is generated inside the battery, the heat transfer liquid absorbs heat from the battery, and then boils and releases heat through vapor generation. The generated vapor can be condensed in an external condenser and then returned to the pool boiling-based battery thermal management system. The examined pool boiling-based battery thermal management system is observed to have relatively higher cooling loads than air-based battery thermal management systems. In addition, the pool boiling-based system is capable of providing a uniform temperature for a battery cell. For instance, Van Gils et al. (2014) observed the temperature difference between the top and bottom of the

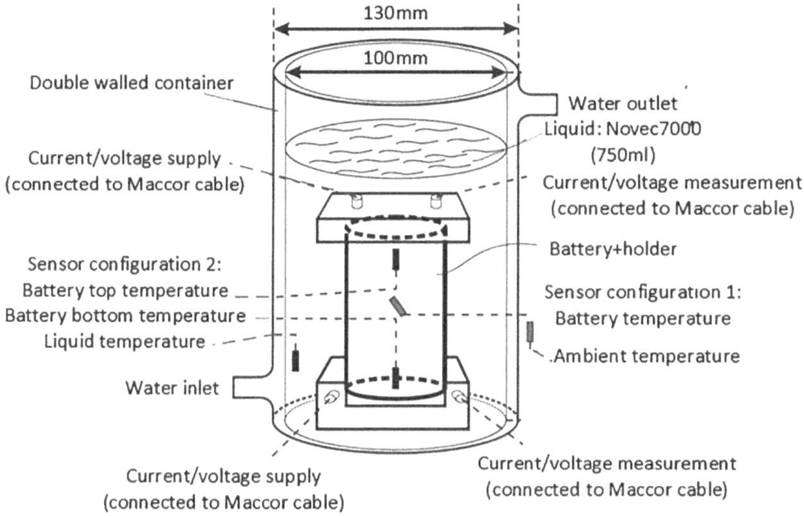

Fig. 5.23 Experimental setup developed by Van Gils et al. (2014) for testing an evaporative pool boing-based battery thermal management system. *(From Van Gils, R.W., Danilov, D., Notten, P.H.L., Speetjens, M.F.M., Nijmeijer, H., 2014. Battery thermal management by boiling heat-transfer. Energy Convers. Manag. 79, 9–17.)*

cylindrical cell to be $0.7°C$ during sensible heat transfer between the liquid heat transfer fluid and the battery. This temperature difference goes to zero when boiling starts. They found that the boiling process is affected by the pressure of the boiling chamber. A decrease in the boiling chamber pressure increases the boiling intensity, resulting in a more rapid response to a battery temperature rise. The authors claim that their pool boiling-based system exhibits better thermal performance, higher cooling load capacity and faster response to battery temperature rise compared to conventional battery thermal management systems.

5.8 Recent developments in battery thermal management systems

In recent years many areas related to different types of battery thermal management systems have been the subject of investigations, assessments and advances. Some of them are discussed in this chapter and some are listed and summarized in Table 5.2. In particular, Table 5.2 summarizes selected recent research on various types of thermal management systems for cylindrical and prismatic batteries. These studies are mainly focused on improving

Table 5.2 Summary of selected recent studies on battery thermal management systems and their key findings.

Battery thermal management system type	Key findings	Source(s)
Air-based, liquid-based, and PCM-based battery thermal management systems	Three types of conventional battery thermal management systems (air-based, liquid-based, and PCM-based) are studied and compared. Improvements in PCM thermal conductivity are discussed	Rao and Wang (2011)
Air-based, liquid-based, and PCM-based battery thermal management systems	A comprehensive review on recent progress, challenges and perspectives of thermal management systems for batteries is presented. This includes thermal models from recent studies which are used to predict heat generation, heat transfer and the temperature distribution within the battery	Lin et al. (2021)
PCM/heat pipe-based	The performance of a battery thermal management system with a PCM and a heat pipe is studied and compared to that of the system solely with a heat pipe. It is shown that a PCM can effectively reduce the temperature difference in the battery pack	Chen et al. (2021)
PCM based	PCM-based battery thermal management system is reviewed for electric devices, photovoltaic modules and batteries. Thermal properties of PCM are discussed	Ling et al. (2014)

Table 5.2 Summary of selected recent studies on battery thermal management systems and their key findings—cont'd

Battery thermal management system type	Key findings	Source(s)
PCM based	Challenges and opportunities are assessed for battery electric vehicles using PCM-based battery thermal management systems. Methods are discussed for improving heat transfer in PCM-based systems	Malik et al. (2016)
PCM based	Methods for improving heat transfer in PCM-based battery thermal management system are discussed	Pan et al. (2016)
Air based	A new cooling design using spoilers in the airflow distribution plenum of a parallel air-cooling system is proposed to improve the performance of a cooling process for a battery pack	Zhang et al. (2021)
Air based	Optimal space between prismatic battery cells is determined using flow resistance network model. Optimal flow pattern and optimal position of inlet and outlet ports of air flow are determined	Chen et al. (2017, 2019)
Air based	Reciprocating air flow for a pack containing cylindrical batteries is examined in terms of improvement in temperature uniformity of battery system	Mahamud and Park (2011)
Air based	Effect of cell arrangement and flow path (e.g., plate angle of plenums) are studied for a set of cylindrical batteries	Liu et al. (2014)

Continued

Table 5.2 Summary of selected recent studies on battery thermal management systems and their key findings—cont'd

Battery thermal management system type	Key findings	Source(s)
Air based	The effect of number and size of cooling channels for cooling prismatic and cylindrical batteries are investigated	Xun et al. (2013)
Air based	Various air flow rates and distances between cells are investigated for a set of cylindrical battery cells	He et al. (2014)
Liquid based	A review of experimental and simulation studies for a liquid-based battery thermal management system for electric vehicles is presented	Kalaf et al. (2021)
Liquid based	Optimization is used to find the optimum geometry of channel (width, position of cooling channels) in terms of pressure drop, temperature uniformity and mean temperature	Jarrett and Kim (2011, 2014)
Liquid based	An oblique mini-channel is proposed for a liquid-based cold plate. The oblique channel prevents a decline in convective heat transfer from inlet to outlet of channel	Jin et al. (2014)
Liquid based	A liquid-based thermal management system is proposed in which heat transfer area from battery to liquid coolant is varied along the flow direction. The system exhibits better temperature uniformity and lower maximum cell temperatures	Rao et al. (2017)

Table 5.2 Summary of selected recent studies on battery thermal management systems and their key findings—cont'd

Battery thermal management system type	Key findings	Source(s)
Liquid based	Several operating conditions are investigated, including rate of charge and discharge (C-rate) on temperature distribution in mini-channel of cooling system and battery surface	Panchal et al. (2017)
Liquid based	A silica-liquid-based cooling plate is proposed consisting of thermal silica plates and copper tubes for a battery thermal management system. Various numbers of silica channels and plates, flow speed and flow paths for several discharge rates are investigated	Wang et al. (2017)
Heat pipe based	A tube-shaped heat pipe with flat evaporator section using water as the heat transfer fluid is examined for cooling prismatic batteries. A thermostat bath supplies cooling for the condensation part of heat pipe. The effect of heating power and angle of the heat pipe on the thermal performance of the system are investigated	Rao et al. (2013)
Heat pipe based	A bent tube heat pipe is used for cooling cylindrical batteries using demineralized water. Forced convection via an external fan is used to condensate the generated demineralized water vapor. The effect of air flow direction and velocity and angle of heat pipe on the	Tran et al. (2014)

Continued

Table 5.2 Summary of selected recent studies on battery thermal management systems and their key findings—cont'd

Battery thermal management system type	Key findings	Source(s)
	thermal performance of the system is studied	
Heat pipe based	An oscillating heat pipe-based thermal management system with acetone as the heat transfer fluid is used for prismatic batteries. A thermostat bath is used for condensation of the acetone vapor. The angle of heat pipe and heating power are investigated on the thermal performance of the system	Wang et al. (2016)
Heat pipe based	An ultra-thin micro heat pipe-based cooling system is applied to a prismatic battery module. Forced convection with air is used to condense the water vapor. The heat pipe arrangement and discharge rate are examined experimentally	Liu et al. (2016)
Pool boiling based	A refrigerant-based battery thermal management system is proposed, consisting of an evaporating pool boiling-based system using stationary R134a as the coolant in which the cylindrical batteries are submerges. Air conditioning of a car condenses the R134a vapor. This system is compared with conventional air- and liquid-based battery thermal management systems	Al-Zareer et al. (2018b)

Table 5.2 Summary of selected recent studies on battery thermal management systems and their key findings—cont'd

Battery thermal management system type	Key findings	Source(s)
Pool boiling based	A battery cooling system is proposed for a hydrogen fueled hybrid electric vehicle. A new aluminum cooling plate design is optimized in terms of uniformity of battery temperature and maximum temperature of battery	Al-Zareer et al. (2018c)
Pool boiling based	The thermal performance of an evaporating pool boiling-based cooling system is compared for various heat transfer fluids (propane, ammonia, R134a refrigerant, hydrogen is performed). The thermal performances of these systems are compared with conventional air- and liquid-based battery thermal management systems	Al-Zareer et al. (2019)
Pool boiling based	A battery thermal management system is proposed based on boiling liquid battery cooling. The batteries are immersed in hydrofluoroether liquid, which has high electric resistance and is noninflammable and environmentally benign. The cooling system is capable of maintaining a battery temperature around its boiling temperature (i.e., 35°C) continuously, even during high rates of discharge and charge	Hirano et al. (2014)

conventional battery thermal management systems (i.e., air-based, liquid-based, and PCM-based) through integration between them, optimization of key parameters and proposing new categories of thermal management systems for batteries.

5.9 Closing remarks

Recent developments in battery systems are making them an increasingly important technology for storage of electrical energy. The operating temperatures of a battery significantly affect its performance, lifetime, and operating safety. Consequently, effective and efficient thermal management systems are needed to control battery temperature by removing heat generated inside a battery or providing heat at cold weather conditions, and providing ventilation when the battery generates exhaust gases during abnormal incidents like thermal runaway. In this chapter, conventional battery thermal management systems, including air-based, liquid-based, and PCM-based systems as well as combinations of these, are described. Also covered are recently proposed new types of battery thermal management systems, such as evaporative pool boiling-based systems.

In air-based battery thermal management systems, the direction of air flow significantly affects the capability of a thermal management system for controlling battery temperature and providing a uniform temperature throughout a battery pack. Therefore, designing new pathways for the air flow through a set of batteries with reasonably small pressure drops has been the focus of much research. Despite of advantages of air-based thermal management systems, including simple design and low power consumption, they are not effective at providing a uniform temperature distribution throughout a battery pack in high-powered battery applications, in part because of the low heat capacity and low thermal conductivity of air.

Liquid-based battery thermal management systems are the most widely used thermal management systems (especially for high-powered battery applications). Liquid-based systems avoid some of the disadvantages of air-based system. In a liquid-based battery thermal management system, the flow direction of the heat transfer fluid is one of the main parameters that affects uniformity of temperature in the battery pack. In addition, the type of liquid heat transfer fluid and the heat transfer surface between the battery and heat transfer fluid (which helps ensure the same rate of heat transfer for all batteries) are important design parameters for liquid-based battery thermal management systems.

PCM-based battery thermal management systems are passive thermal management technologies that, unlike liquid-based systems which consume considerable power, can store and/or release large amounts of thermal energy without additional energy consumption. However, this type of battery thermal management system can provide good temperature uniformity (since the latent heat is used to remove heat from a battery pack rather than sensible heat). But they are inadequate for the continuous operation required to lower effectively the temperature difference in a battery pack, in part due to the low thermal conductivity of PCMs. The thermal conductivity of the PCM used in PCM-based systems can be enhanced through integration of heat pipe and air and liquid-based cooling systems and adding some materials with high thermal conductivities to the PCM.

The disadvantages of the low thermal conductivity of the PCMs typically used in PCM-based battery thermal management systems are offset in liquid-vapor phase change-based battery thermal management systems that include heat pipe-based and evaporative pool boiling-based systems. Specifically, the recently proposed evaporative pool boiling-based battery thermal management system is a promising method for removing heat from batteries. Pool boiling-based battery thermal management systems are capable of supplying higher cooling loads for batteries than conventional cooling systems, without the need for an external power supplier device. In such systems, the batteries are fully or partially submerged in a stationary heat transfer fluid. The temperature of the heat transfer fluid increases as it absorbs battery heat initially, and then remains constant during the vaporization of the heat transfer fluid as further heat is taken in. This relatively new category of battery thermal management system has a large thermal capacity (mainly due to latent heat of heat transfer fluid) and a high thermal conductivity, which can cool batteries more effectively than conventional thermal management systems.

Study questions

5.1. What are the main steps in the design of a battery thermal management system?

5.2. Consider the cabin air cooling system shown in Fig. 5.24, which uses a vapor compression refrigeration cycle to cool a battery pack. How does this system dissipate heat from the battery pack? What are the

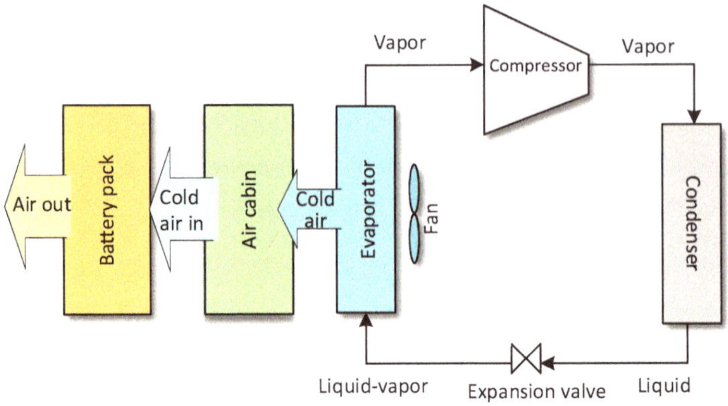

Fig. 5.24 Cabin air cooling system using vapor compression refrigeration cycle to cool a battery pack.

advantages and disadvantages of this battery thermal management system?

5.3. Compare air-based and liquid-based battery thermal management systems.

5.4. Compare liquid-based and phase change material (PCM)-based battery thermal management systems.

5.5. What are direct and indirect liquid-based battery thermal management systems? List examples for each of them.

5.6. Consider the battery thermal management system shown in Fig. 5.25. What is the category of this thermal management system? How does this system cool the battery pack? Explain.

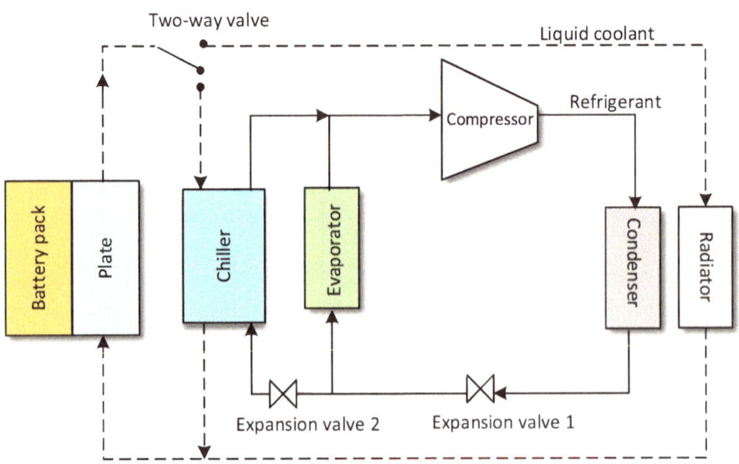

Fig. 5.25 Battery thermal management system for cooling a battery pack.

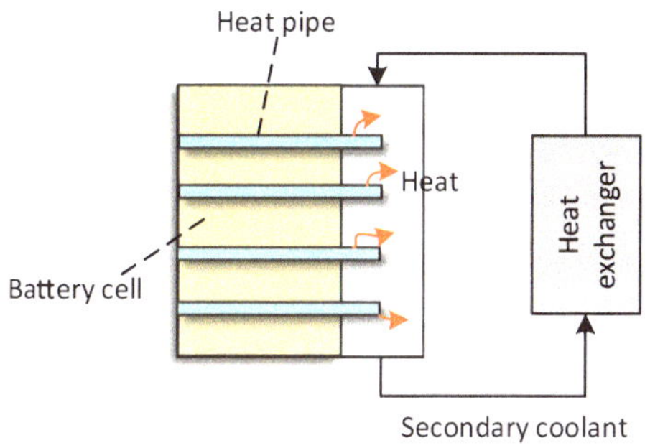

Fig. 5.26 Heat pipe-based battery thermal management system with its secondary cooling system.

5.7. Compare the temperature profile with time for PCM-based and air/liquid-based battery thermal management systems. Which type of these battery thermal management systems has the higher heat adsorption capacity? Why?

5.8. Consider the heat pipe-based battery thermal management system shown in Fig. 5.26. How does this system remove heat from the battery pack? Explain.

5.9. Consider pool boiling-based and air-based battery thermal management systems used to maintain the temperature of battery within a desired operating range. Which of these two systems is expected to provide a more uniform temperature throughout the battery cells in the pack? Why?

References

Al-Zareer, M., Dincer, I., Rosen, M.A., 2017a. Electrochemical modeling and performance evaluation of a new ammonia-based battery thermal management system for electric and hybrid electric vehicles. Electrochim. Acta 247, 171–182.

Al-Zareer, M., Dincer, I., Rosen, M.A., 2017b. Novel thermal management system using boiling cooling for high-powered lithium-ion battery packs for hybrid electric vehicles. J. Power Sources 363, 291–303.

Al-Zareer, M., Dincer, I., Rosen, M.A., 2018a. A review of novel thermal management systems for batteries. Int. J. Energy Res. 42 (10), 3182–3205.

Al-Zareer, M., Dincer, I., Rosen, M.A., 2018b. Heat and mass transfer modeling and assessment of a new battery cooling system. Int. J. Heat Mass Transf. 126, 765–778.

Al-Zareer, M., Dincer, I., Rosen, M.A., 2018c. Performance assessment of a new hydrogen cooled prismatic battery pack arrangement for hydrogen hybrid electric vehicles. Energy Convers. Manag. 173, 303–319.

Al-Zareer, M., Dincer, I., Rosen, M.A., 2019. Comparative assessment of new liquid-to-vapor type battery cooling systems. Energy 188, 116010.

Basu, S., Hariharan, K.S., Kolake, S.M., Song, T., Sohn, D.K., Yeo, T., 2016. Coupled electrochemical thermal modelling of a novel Li-ion battery pack thermal management system. Appl. Energy 181, 1–13.
Chen, K., Hou, J., Song, M., Wang, S., Wu, W., Zhang, Y., 2021. Design of battery thermal management system based on phase change material and heat pipe. Appl. Therm. Eng. 188, 116665.
Chen, K., Wang, S., Song, M., Chen, L., 2017. Configuration optimization of battery pack in parallel air-cooled battery thermal management system using an optimization strategy. Appl. Therm. Eng. 123, 177–186.
Chen, K., Wu, W., Yuan, F., Chen, L., Wang, S., 2019. Cooling efficiency improvement of air-cooled battery thermal management system through designing the flow pattern. Energy 167, 781–790.
Duan, X., Naterer, G.F., 2010. Heat transfer in phase change materials for thermal management of electric vehicle battery modules. Int. J. Heat Mass Transf. 53 (23-24), 5176–5182.
Goli, P., Legedza, S., Dhar, A., Salgado, R., Renteria, J., Balandin, A.A., 2014. Graphene-enhanced hybrid phase change materials for thermal management of Li-ion batteries. J. Power Sources 248, 37–43.
He, F., Li, X., Ma, L., 2014. Combined experimental and numerical study of thermal management of battery module consisting of multiple Li-ion cells. Int. J. Heat Mass Transf. 72, 622–629.
Hémery, C.V., Pra, F., Robin, J.F., Marty, P., 2014. Experimental performances of a battery thermal management system using a phase change material. J. Power Sources 270, 349–358.
Hirano, H., Tajima, T., Hasegawa, T., Sekiguchi, T., Uchino, M., 2014. Boiling liquid battery cooling for electric vehicle. In: 2014 IEEE Conference and Expo Transportation Electrification Asia-Pacific (ITEC Asia-Pacific), Beijing, China, pp. 1–4.
Hong, S., Zhang, X., Chen, K., Wang, S., 2018. Design of flow configuration for parallel air-cooled battery thermal management system with secondary vent. Int. J. Heat Mass Transf. 116, 1204–1212.
Huang, C., Wang, Q., Rao, Z., 2015. Thermal conductivity prediction of copper hollow nanowire. Int. J. Therm. Sci. 94, 90–95.
Jaguemont, J., Omar, N., Van den Bossche, P., Mierlo, J., 2018. Phase-change materials (PCM) for automotive applications: a review. Appl. Therm. Eng. 132, 308–320.
Jarrett, A., Kim, I.Y., 2011. Design optimization of electric vehicle battery cooling plates for thermal performance. J. Power Sources 196 (23), 10359–10368.
Jarrett, A., Kim, I.Y., 2014. Influence of operating conditions on the optimum design of electric vehicle battery cooling plates. J. Power Sources 245, 644–655.
Javani, N., Dincer, I., Naterer, G.F., Yilbas, B.S., 2014. Heat transfer and thermal management with PCMs in a Li-ion battery cell for electric vehicles. Int. J. Heat Mass Transf. 72, 690–703.
Jiang, G., Huang, J., Liu, M., Cao, M., 2017. Experiment and simulation of thermal management for a tube-shell Li-ion battery pack with composite phase change material. Appl. Therm. Eng. 120, 1–9.
Jin, L.W., Lee, P.S., Kong, X.X., Fan, Y., Chou, S.K., 2014. Ultra-thin minichannel LCP for EV battery thermal management. Appl. Energy 113, 1786–1794.
Kalaf, O., Solyali, D., Asmael, M., Zeeshan, Q., Safaei, B., Askir, A., 2021. Experimental and simulation study of liquid coolant battery thermal management system for electric vehicles: a review. Int. J. Energy Res. 45 (5), 6495–6517.
Kandasamy, R., Wang, X.Q., Mujumdar, A.S., 2007. Application of phase change materials in thermal management of electronics. Appl. Therm. Eng. 27 (17-18), 2822–2832.
Khan, M.R., Swierczynski, M.J., Kær, S.K., 2017. Towards an ultimate battery thermal management system: a review. Batteries 3 (1), 9.

Li, W.Q., Qu, Z.G., He, Y.L., Tao, Y.B., 2014. Experimental study of a passive thermal management system for high-powered lithium ion batteries using porous metal foam saturated with phase change materials. J. Power Sources 255, 9–15.

Lin, J., Liu, X., Li, S., Zhang, C., Yang, S., 2021. A review on recent progress, challenges and perspective of battery thermal management system. Int. J. Heat Mass Transf. 167, 120834.

Ling, Z., Wang, F., Fang, X., Gao, X., Zhang, Z., 2015. A hybrid thermal management system for lithium ion batteries combining phase change materials with forced-air cooling. Appl. Energy 148, 403–409.

Ling, Z., Zhang, Z., Shi, G., Fang, X., Wang, L., Gao, X., Fang, Y., Xu, T., Wang, S., Liu, X., 2014. Review on thermal management systems using phase change materials for electronic components, Li-ion batteries and photovoltaic modules. Renew. Sust. Energ. Rev. 31, 427–438.

Liu, F., Lan, F., Chen, J., 2016. Dynamic thermal characteristics of heat pipe via segmented thermal resistance model for electric vehicle battery cooling. J. Power Sources 321, 57–70.

Liu, Z., Wang, Y., Zhang, J., Liu, Z., 2014. Shortcut computation for the thermal management of a large air-cooled battery pack. Appl. Therm. Eng. 66 (1-2), 445–452.

Mahamud, R., Park, C., 2011. Reciprocating air flow for Li-ion battery thermal management to improve temperature uniformity. J. Power Sources 196 (13), 5685–5696.

Malik, M., Dincer, I., Rosen, M.A., 2016. Review on use of phase change materials in battery thermal management for electric and hybrid electric vehicles. Int. J. Energy Res. 40 (8), 1011–1031.

Mehrali, M., Latibari, S.T., Rosen, M.A., Akhiani, A.R., Naghavi, M.S., Sadeghinezhad, E., Metselaar, H.S.C., Nejad, M.M., Mehrali, M., 2016. From rice husk to high performance shape stabilized phase change materials for thermal energy storage. R Soc. Chem. Adv. 6 (51), 45595–45604.

Pan, D., Xu, S., Lin, C., Chang, G., 2016. Thermal management of power batteries for electric vehicles using phase change materials: a review. SAE Technical Paper 2016, 13.

Pesaran, A., Keyser, M., Burch, S., 1999. An approach for designing thermal management systems for electric and hybrid vehicle battery packs. Report No. NREL/CP-540-25992, National Renewable Energy Laboratory, Golden (US).

Panchal, S., Khasow, R., Dincer, I., Agelin-Chaab, M., Fraser, R., Fowler, M., 2017. Thermal design and simulation of mini-channel cold plate for water cooled large sized prismatic lithium ion battery. Appl. Therm. Eng. 122, 80–90.

Rao, Z., Qian, Z., Kuang, Y., Li, Y., 2017. Thermal performance of liquid cooling based thermal management system for cylindrical lithium-ion battery module with variable contact surface. Appl. Therm. Eng. 123, 1514–1522.

Rao, Z., Wang, S., 2011. A review of power battery thermal energy management. Renew. Sust. Energ. Rev. 15 (9), 4554–4571.

Rao, Z., Wang, S., Wu, M., Lin, Z., Li, F., 2013. Experimental investigation on thermal management of electric vehicle battery with heat pipe. Energy Convers. Manag. 65, 92–97.

Sabbah, R., Kizilel, R., Selman, J.R., Al-Hallaj, S., 2008. Active (air-cooled) vs. passive (phase change material) thermal management of high power lithium-ion packs: limitation of temperature rise and uniformity of temperature distribution. J. Power Sources 182 (2), 630–638.

Santhanagopalan, S., Smith, K., Neubauer, J., Kim, G.H., Keyser, M., Pesaran, A., 2014. Design and Analysis of Large Lithium-Ion Battery Systems. Artech House.

Shirazi, A.H.N., Mohebbi, F., Azadi Kakavand, M.R., He, B., Rabczuk, T., 2016. Paraffin nanocomposites for heat management of lithium-ion batteries: a computational investigation. J. Nanomater. 2016, 10.

Sun, H., Dixon, R., 2014. Development of cooling strategy for an air cooled lithium-ion battery pack. J. Power Sources 272, 404–414.

Tran, T.H., Harmand, S., Sahut, B., 2014. Experimental investigation on heat pipe cooling for hybrid electric vehicle and electric vehicle lithium-ion battery. J. Power Sources 265, 262–272.
Van Gils, R.W., Danilov, D., Notten, P.H.L., Speetjens, M.F.M., Nijmeijer, H., 2014. Battery thermal management by boiling heat-transfer. Energy Convers. Manag. 79, 9–17.
Wang, C., Zhang, G., Meng, L., Li, X., Situ, W., Lv, Y., Rao, M., 2017. Liquid cooling based on thermal silica plate for battery thermal management system. Int. J. Energy Res. 41 (15), 2468–2479.
Wang, Q., Rao, Z., Huo, Y., Wang, S., 2016. Thermal performance of phase change material/oscillating heat pipe-based battery thermal management system. Int. J. Therm. Sci. 102, 9–16.
Wang, T., Tseng, K.J., Zhao, J., Wei, Z., 2014. Thermal investigation of lithium-ion battery module with different cell arrangement structures and forced air-cooling strategies. Appl. Energy 134, 229–238.
Wu, W., Wu, W., Wang, S., 2019. Form-stable and thermally induced flexible composite phase change material for thermal energy storage and thermal management applications. Appl. Energy 236, 10–21.
Wu, W., Wu, W., Wang, S., 2017a. Thermal optimization of composite PCM based large-format lithium-ion battery modules under extreme operating conditions. Energy Convers. Manag. 153, 22–33.
Wu, W., Yang, X., Zhang, G., Chen, K., Wang, S., 2017b. Experimental investigation on the thermal performance of heat pipe-assisted phase change material based battery thermal management system. Energy Convers. Manag. 138, 486–492.
Wu, W., Yang, X., Zhang, G., Ke, X., Wang, Z., Situ, W., Li, X., Zhang, J., 2016. An experimental study of thermal management system using copper mesh-enhanced composite phase change materials for power battery pack. Energy 113, 909–916.
Xu, J., Lan, C., Qiao, Y., Ma, Y., 2017. Prevent thermal runaway of lithium-ion batteries with minichannel cooling. Appl. Therm. Eng. 110, 883–890.
Xu, X.M., He, R., 2013. Research on the heat dissipation performance of battery pack based on forced air cooling. J. Power Sources 240, 33–41.
Xun, J., Liu, R., Jiao, K., 2013. Numerical and analytical modeling of lithium ion battery thermal behaviors with different cooling designs. J. Power Sources 233, 47–61.
Yang, T., Yang, N., Zhang, X., Li, G., 2016b. Investigation of the thermal performance of axial-flow air cooling for the lithium-ion battery pack. Int. J. Therm. Sci. 108, 132–144.
Yang, X.H., Tan, S.C., Liu, J., 2016a. Thermal management of Li-ion battery with liquid metal. Energy Convers. Manag. 117, 577–585.
Zhang, F., Lin, A., Wang, P., Liu, P., 2021. Optimization design of a parallel air-cooled battery thermal management system with spoilers. Appl. Therm. Eng. 182, 116062.
Zhang, P., Hu, Y., Song, L., Ni, J., Xing, W., Wang, J., 2010. Effect of expanded graphite on properties of high-density polyethylene/paraffin composite with intumescent flame retardant as a shape-stabilized phase change material. Sol. Energy Mater. Sol. Cells 94 (2), 360–365.
Zhao, J., Lv, P., Rao, Z., 2017. Experimental study on the thermal management performance of phase change material coupled with heat pipe for cylindrical power battery pack. Exp. Thermal Fluid Sci. 82, 182–188.
Zhou, X, Xiao, H, Feng, J, Zhang, C, Jiang, Y, 2009. Preparation and thermal properties of paraffin/porous silica ceramic composite. Compos. Sci. Technol. 69 (7–8), 1246–1249.

CHAPTER 6

Battery system design

Chapter objectives

- Present and explain the main process steps in the design of a battery system.
- Describe the design of a battery management system, as well as electrical, mechanical, and thermal designs for a battery pack.
- Discuss the optimum design of a battery pack that can extend the lifetime of the battery cells and provide safe and reliable operation of the battery pack.
- Provide insights into important factors in the design of a battery pack.
- Discuss the layout of a battery management system and its potential interference with other subsystems.
- Present and explain various arrangements of battery cells in a pack.
- Describe a stress-strain analysis and an assessment of vibrations for the structure of a battery pack.
- Explain and discuss the selection of a suitable battery thermal management system, accounting for its application.

Nomenclature

A area (m^2)
C cost ($)
E energy (kJ)
F force (N)
h heat transfer coefficient (W/m^2 K)
I current (A)
k thermal conductivity (W/m K)
L length (m)
N_B number of battery cells
q heat flux (W/m^2)
Q battery capacity (Ah)
$\dot{Q}$ heat rate (W)
R electrical resistance (Ω), thermal resistance (K/W, °C/W)
T temperature (°C, K)
t thickness (m)
V voltage (V)
Y modulus of elasticity, Young's modulus (kPa)

Battery Technology
https://doi.org/10.1016/B978-0-443-18862-6.00006-9

Greek letters

σ stress (kPa)
ε strain
ρ density (kg/m^3)
J Joule
σ Stefan-Boltzmann constant (5.670367×10^{-8} $kg/(s^3\ K^4)$)
ε surface emissivity

Subscripts

a	ambient
B	battery
cond	conduction
conv	convection
nom	nominal
Rad	radiation
T	total

6.1 Introduction

Battery energy storage systems are currently used for a wide range of applications with low to high energy storage capacities. The number of cells in battery systems can be varied from a single cell to a few cells for electronic devices, as well as from tens to thousands of battery cells for relatively large energy capacities such as required for automotive applications. Understanding the application helps in identifying the requirements for a battery system. Table 6.1 presents a typical range of required voltages for a battery system for various applications. The design of a battery system depends on the type of application, electrochemical technology used, design of electrochemical cell, as well as safety, life, and thermal characterization.

One of the first tasks in battery pack design is to define the technology and number of cells to use and to meet the technical requirements including desired voltage, power and energy. Several factors such as cost, safety, lifetime, compatibility of cell for packaging and power to energy ratio of the cell for a given application are considered during the selection of a cell when designing a battery energy storage system. The power to energy ratio indicates the duration that the battery cell can operate while delivering its power output. It is also known as discharge time of the battery. The power to energy ratio is determined by dividing the energy capacity

Table 6.1 Typical voltage range required for a battery system for various applications.

Application	Voltage range (V)	Energy storage capacity range (Wh)	Battery type
Small-scale applications (e.g., phones, laptops)	48–72	100–15,000	Prismatic or cylindrical
Medium-scale applications (e.g., electric/hybrid electric vehicles)	320–360	20,000–70,000	Cylindrical
Large-scale applications (e.g., electric/hybrid electric buses and trucks)	720–760	60,000–300,000	Cylindrical

Data in this table are drawn by the present authors from many articles and websites, and represent typical ranges for various applications.

by the electrical power. In this chapter, it is assumed that the electrochemical cell has been wisely selected and the battery pack design is the main issue addressed throughout this chapter.

A typical battery system generally includes a number of cells arranged in a pack. These terms are central to the chapter and can be described as follows:

- Cell: A cell is the basic unit of a battery energy storage system. It is where the electrochemical reactions take place through which electrical energy is converted to chemical energy during charge and converted back to the electrical energy during discharge. A battery cell is basically comprised of an anode, a cathode, a separator and an electrolyte. The cell chemistry (and material used in the cell) is one of the main contributors to the cost of a battery system. The materials used in fabrication of battery cells can be nickel, iron, cadmium, lithium, etc. depending on the type of battery. The quantity of material used in a battery cell depends on the energy delivered, typically in watt-hour (Wh), per kg of material used.
- Pack: A battery pack is an assembly containing a number of cells, in a frame. The frame is designed to protect the battery pack and its cells from external shocks, vibrations and heat. A battery pack contains an arrangement of modules, as well as various protection/control systems. These include systems for meeting thermal and mechanical requirements and battery management, and they typically are embedded in the pack design. Arrangements of battery cells in a pack is an important design factors and is discussed in detail in Section 6.3.

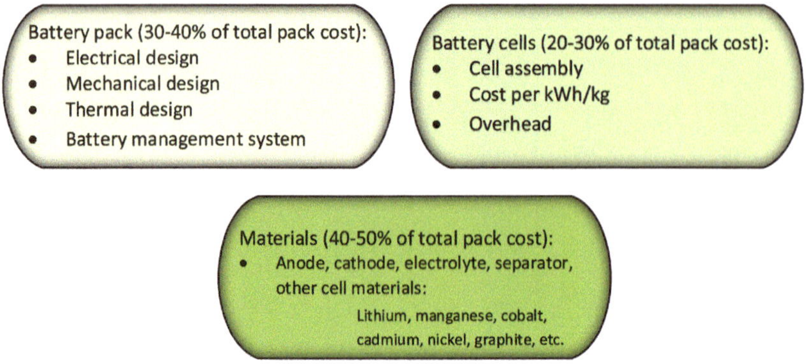

Fig. 6.1 Primary cost contributors in the design of a battery pack.

Fig. 6.1 shows typical cost distributions and required elements for designing a battery storage system for an electric vehicle. It is seen that the cell chemistry has the highest cost contribution for designing a battery energy storage system. However, this value can be varied based on the type of material used and the corresponding battery chemistry.

Generally, battery pack design involves determining the number of cells assembled to form a battery pack for a required voltage and capacity. The battery storage system design is analogous to the cell design in that there is a trade-off between performance, uniform current distribution between cells during charge and discharge, safety, and cost of each step of the process design. The main process steps for design of a battery system include electrical, thermal, and mechanical design, as well as battery management design, along with addressing of safety and cost considerations. Fig. 6.2 shows the main process designs for a battery pack. The primary aim of *electrical design* is to predict the electrical power that the battery should provide, as well as to provide a balanced current/voltage distribution among the battery cells. The ultimate goal of *thermal design* is to maintain the battery system within a uniform and constant desirable temperature by removing heat generated within the battery cells. The *mechanical design* is generally aimed at applying the proper pressure to the battery system to avoid bulges in the cells, to protect the battery cells from vibrations, and to address other safety considerations. The *battery management system* is primarily used in the design of a battery system to monitor and maintain the current, voltage and temperature of the batteries during the operation of system. When any of these operating

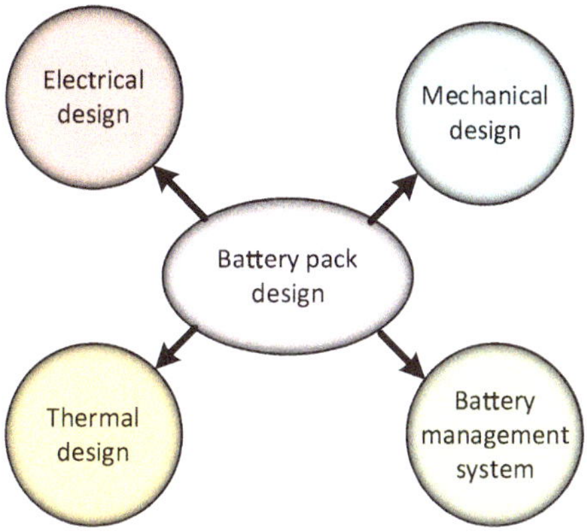

Fig. 6.2 Primary design processes for development of a battery pack.

parameters exceed from their limiting values, the management system modifies or halts the operation of the battery system.

This chapter provides insights into important points in these design areas such as the layout of the battery management system and its interference with other subsystems, arrangements of battery cells in the pack, stress-strain analysis and vibrations in the structure of the pack, and selection of suitable battery management system accounting for the application (e.g., electric/hybrid electric vehicle). In the remainder of the chapter, all of these design processes for a battery energy storage system are described and discussed in detail, including battery management system design as well as electrical, mechanical, and thermal design.

6.2 Battery management system

A battery management system typically is an electronic control unit that regulates and monitors the operation of a battery during charge and discharge. In addition, the battery management system is responsible for connecting with other electronic units and exchanging the necessary data about battery parameters. The voltage, capacity, temperature, power consumption, state of charge and health, charging cycle, and other characteristics of the battery are controlled and monitored by the battery management system. The battery management system uses these data for estimation of the state of charge and state of health of a battery pack. The battery *state of health* is an indicator for the present battery capacity compared to the capacity at the beginning of its life. This indicator can be used to predict the end of life of a battery.

The *state of charge* (SOC) represents the ratio of available capacity at a given time to the maximum possible charge that can be stored in the battery. Each battery should be maintained between the minimum and maximum state of charge defined in the battery profile.

The operation of a battery management system is generally aimed at an optimum utilization of stored energy in the battery cells. The battery management system protects the battery cells from deep discharge and overcharge, which respectively result from extremely high discharge and fast charge of batteries. Fig. 6.3 shows a battery management system coupled with a battery pack for optimum and safe operation of the battery energy storage system in an electric vehicle. A controller area network (CAN bus) is a robust vehicle bus standard designed to allow processors like microcontrollers and other devices to communicate with each other. It enables multiple battery cells to operate as one system for data measurement (voltage V, current I, temperature T, etc.).

Another main task of a battery management system is a cell balancing function through which the same discharge and charge requirements for each battery cell are provided. By cell balancing, the cells are maintained at equal voltage levels and the capacity utilization of the battery pack is maximized. This can be attained through the use of *power metal-oxide-semiconductor field-effect transistor* (*MOSFET*) in the battery management system. The power

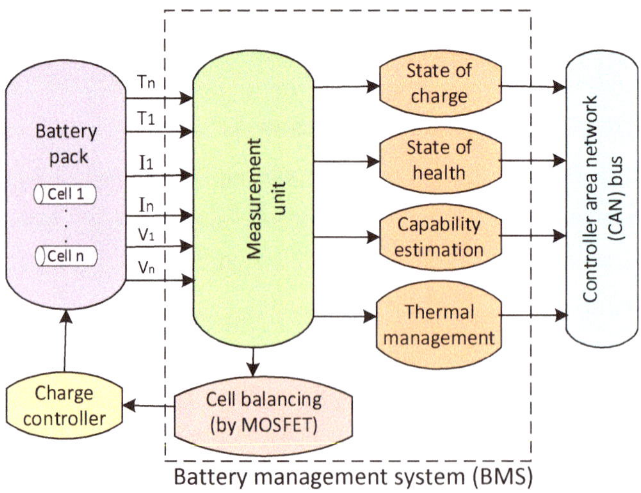

Fig. 6.3 Battery management system (BMS) coupled with a battery pack in an electric vehicle.

MOSFET is a special metal oxide semiconductor field effect transistor and a type of multicarrier electric control unit that is designed for processing various power loads. The power MOSFET is capable of controlling current and voltage output, which provides a stable and safe condition for the battery system (see Fig. 6.3). The power MOSFET is straightforward to operate and its maintenance is easy, making it the most widely used unit in numerous applications of battery systems.

Fig. 6.4 shows the main parts of a battery management system. It is seen that one of the core functions of a battery management system is to check and monitor the temperature, voltage and current, so as to maintain them within the specified limits for safe operation of the battery pack.

In addition, a battery management system measures and stores various parameters including cell parameters (open circuit voltage, state of charge, etc.), cell balancing status and input/output requests. Then, based on this information, the battery management system controls switching devices for allowing or preventing operation of one or more subsystems. For example, when the motor requests current and the contactors are open, current cannot flow. So, the battery management system closes the contactors to let current flows. The data are stored in the data storage memory, from which the data can be retrieved and used for calculation of pack parameters.

The battery management system is usually capable of self-diagnosis. During self-diagnosis, the battery management system carries out several tests to check whether all functions and sensors are working properly or not. It also handles the errors and decides about continuing operation or not.

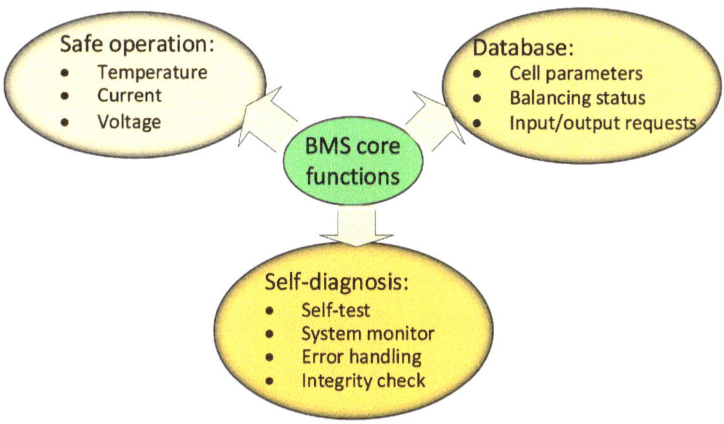

Fig. 6.4 Core functions of a battery management system in a battery pack.

For instance, when errors in the voltage limit that result from the tests are minor and do not lead to safety issues, the battery management system allows for continuing operation of the battery pack.

Several interface drivers (hardware) are placed in the battery management system like a microcontroller unit (MCU) and an integrated circuit (IC) chip which are controlled by software. The MCU receives inputs from switches, sensors, and similar components, and controls the electric motors and displays for a given application (e.g., electric vehicle) in accordance with a preset program that tells it what to do and how to respond. In battery management systems, in addition to the primary MCU, there is secondary processor unit that acts as a secondary MCU in case of failure in the operation of the primary processor.

In addition, an insolation monitoring device (IMD) is often used in the battery management system which is intended for monitoring the insulation resistance and the voltage of a battery during its assembly from individual battery cells. This device is especially useful during high voltage applications of the battery pack as it separates the low voltage line from high voltage. If there is a voltage leakage specifically in high voltage applications, the IMD detects this leakage and signals to the battery management system to cut off the flow of current. Furthermore, current sensors can be used in the battery management system to measure and provide real-time data on the current flowing through the battery. From this data, the state of charge can be determined.

Some typical key tasks of a battery management system follow:

- During charging, if an unequal current flows between the cells, the battery management system equalizes it through monitoring and balancing the voltage and temperature of each module.
- During charging, if a module does not sufficiently receive current, the battery management system bypasses other module(s) to provide more current flow for that module.
- If the temperature of any module rises, the battery management system slows the charge or discharge rate of that module.
- When a module is overcharged, the battery management system can either partially discharge the module with higher voltage through a resistor so that some current will flow to other modules (known as passive balancing) or stop charging the module with higher voltage and instead charge other modules with lower voltages (known as active balancing).
- The battery management system can limit the temperature of modules if active cooling or heating processes are used in the battery storage system.

As stated before, a battery management system has electronic switches that disconnect the battery from a load or charger during critical and dangerous conditions. In other words, the battery management system protects battery from possible damage associated with the cells within the pack. Failure in the operation of a battery management system may result in serious problems, including the following:

- Thermal runaway: Thermal runaway in a battery happens when the temperature of the cell exceeds the onset temperature, the temperature at which battery self-heating commences. With further increases in cell temperature, the degradation reactions occur at a faster rate, eventually emitting various gases and leading to ignition and combustion of the battery cell. The conditions that lead to thermal runaway can be internal short circuits between electrodes, between current collectors and between the anode and the collector.
- Cell capacity decline: Imbalanced current and voltage distributions through the cells results in an increase in the deterioration process rate in the cells, which subsequently leads to capacity fade within the battery cells.
- Death of cell: The wrong determination of state of charge and state of health by the battery management system leads to overcharging or discharge of the battery above or below its specified thresholds and consequently death of the cell.
- Damage of load unit: The load unit can be damaged when there is an imbalanced current distribution through the cells.
- Power MOSFET overheating: Power MOSFET is a heat dissipating device. When heat is not sufficiently removed from power MOSFET, its temperature increases and it becomes overheated. In this condition, the power MOSFET should be replaced or battery pack failure can result.
- Temperature sensor failure: This is a risky condition in which the power MOSFET, as well as the temperatures of the battery cells, is not correctly displayed. This condition may result in battery overheating, melting of cells and thermal runaway. Thus, selection of appropriate temperature sensors with high quality is necessary in battery management systems.

Fig. 6.5 shows a typical battery pack layout in an electrical vehicle. It includes contactors which are connected to the pack outlet, which can be an electrical motor or an electrical motor controller. A contractor is special type of relay used for switching an electrical circuit on or off. It generates a magnetic force that mechanically operates an electric contact. The precharge

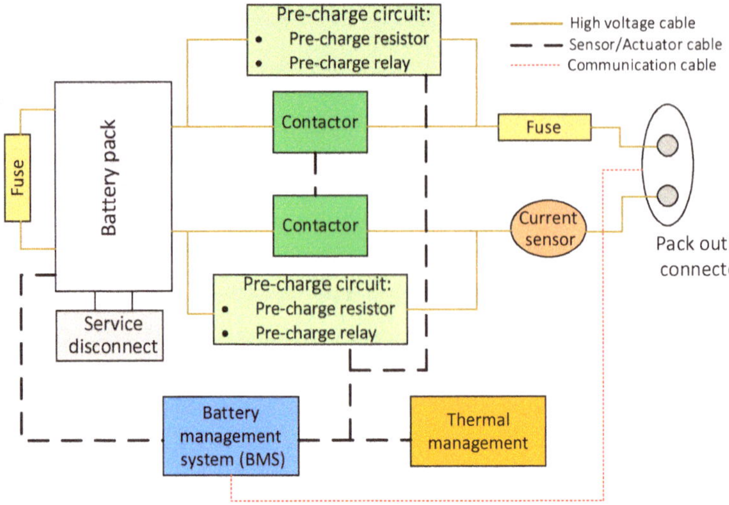

Fig. 6.5 Typical configuration of battery pack components in an electrical vehicle.

circuits are built in the contactors to allow the current to flow in a controlled manner until the voltage level rises to very near the source voltage before the main contactors are permitted to close. Inside the battery pack, the battery cells are arranged for delivering target power. Several fuses are also used in the battery pack and contactors to protect battery cells from potentially damaging and dangerous overcurrent and overcharging events. There is also a thermal management system which is connected to the battery management system.

6.3 Electrical design of battery pack

One of the first steps in the electrical design of a battery energy storage system is to determine or predict the electrical power against the time duty cycle to which the battery will be subject. For instance, for an energy intense application, the level of electrical power is low while the charge and discharge durations are long. Also, for intense power applications, the level of electrical power is high whereas the charge and discharge durations are relatively short.

Battery systems are typically sized for higher energy capacities than the required capacity to compensate the capacity fade due to the degradation process. As stated in Section 6.1, determination of the desired power to energy ratio for the electrochemical cell is one of the first key steps in designing a battery system for a given application. In addition to power to energy ratio, several other electrical attributes need to be considered for cell design. These include nominal, maximum, and minimum cell voltages, capacity and thermal behavior/requirements. In the following subsection, the configuration of cells within a battery pack is described.

6.3.1 Configuration of cells within a battery pack

The nominal, maximum, and minimum voltages for a battery pack, as well as the battery capacity, need to be matched to the drive-train unit to which the battery pack is connected.

The arrangement of battery cells in a pack can be in parallel or in series. In the parallel arrangement, the output current and capacity of the pack increases due to an increase in the total number of electrons flowing through the batteries. For a given circuit, from each battery cell a set number of electrons can be drawn per second, so if two or more battery cells are connected in parallel, the number of electrons they can transport per second is determined by multiplication and, subsequently, the total current in the circuit is increased. In the series arrangement of battery cells, the output voltage of the pack increases as the force to move the electrons through the battery cells increases.

Furthermore, battery cells can be connected in a series-parallel arrangement to increase the voltage and current of the battery system simultaneously. The total current I_T and the total nominal voltage $V_{T,\mathrm{nom}}$ for a series-parallel arrangement can be determined as follows:

$$I_T = \sum_n I_n \tag{6.1}$$

$$V_{T,\mathrm{nom}} = \sum_m V_{m,\mathrm{nom}} \tag{6.2}$$

where n denotes number of rows and m number of columns. The rows are connected in series and the columns in parallel. Also, $V_{m,\mathrm{nom}}$ denotes the nominal cell voltage. Similarly, the maximum and minimum pack voltage can be determined based on the single cell maximum and minimum voltage. The total energy to be supplied by the battery pack can be expressed as:

$$E_{\mathrm{pack},T} = V_{T,\mathrm{nom}} Q_T \tag{6.3}$$

where Q_T is the battery capacity (in Ah).

There are two main parallel/series arrangements in a battery module and pack:

- Series-parallel (SP) configuration: In this configuration, the cells are firstly connected in series, and then the series columns are connected in parallel as shown in Fig. 6.6A. An advantage of this configuration is that any series rows can be electrically removed from the battery system in case of a failure in that row. Despite of this advantage, there is a significant concern regarding the use of the SP configuration in some applications like electric vehicles. This concern is related to the nonuniform distribution of voltage between the cells, which results in an internal flow of current among the cells for balancing them. Even with the use of the same battery cells with the same nominal voltage, after a period of battery system operation, a voltage deviation is likely to develop between the cells. With internal flow of current between the cells, the cells experience instant charges and discharges, which damage the cell and reduce the battery lifetime. So, the SP configuration is not desirable for many battery storage system applications, specifically when a large number of battery cells is required.
- Parallel-series (PS) configuration: In this configuration, the cells are connected in parallel to form *modules*, and then parallel rows are connected in

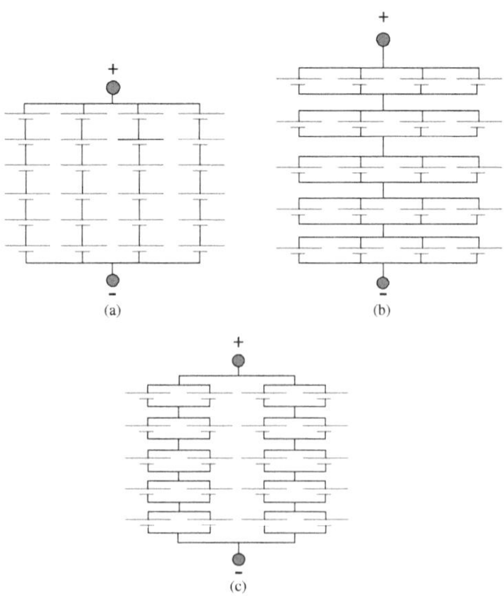

Fig. 6.6 Various arrangement of battery cells in a pack: (A) series-parallel (SP), (B) parallel-series (PS), (C) parallel-series-parallel (PSP).

series (see Fig. 6.6B) to produce a higher voltage. Unlike the SP configuration, the PS configuration benefits from creating circuitry balanced connections through the battery system because of the relatively lower electrical resistance compared to the SP design. Also, the PS configuration provides the capability of battery system operation with partial functionality in the case of a single cell failure. Therefore, the desired electrical design of a battery system is basically to join cells to modules and then modules to packs. In such a design, cells are connected in parallel (with metal bars) to form the modules placed in a metal plate, and then the modules are placed in series to create a battery pack.

In addition, parallel-series-parallel (PSP) connections can be integrated at the same time to form another configuration which is shown in Fig. 6.6C. This configuration is often used for extreme large scale applications and has higher cost and complexity compared to PS and SP configurations. There are some factors which determine the priority of these configurations in terms of trade-off between the complexity of electrical wiring, cost, capability to operate with partial functionality in case of a cell failure, and safety issues.

6.3.1.1 Electrical failure in a battery pack

Generally, when there are voltage/current imbalances in a battery pack system, it usually indicates a poor electrical design of the pack. Continuous imbalances in the battery pack deteriorate the storage capacity and lower the lifetimes of the battery cells. Thus, the electrical design for a battery storage system should be effectively carried out.

It should be noted that the selection of suitable conductor connecting cells within the pack is important, especially when high currents are involved in the battery storage system. Resistance in the conductor leads to a voltage drop in the battery pack. Even a small resistance difference between two electrical paths can result in differences in the voltage drop throughout the pack. This creates an imbalance in the current distributions between the cells, as some cells receive more current than others. In addition, a single connection between sequential modules in the battery pack creates different paths in which more current is passed through a shorter distance which has smaller resistance. Thus, to create a uniform current distribution among the cells in a single module of a PS configuration, multiple connections between the modules are suggested rather than a single connection (see Fig. 6.7).

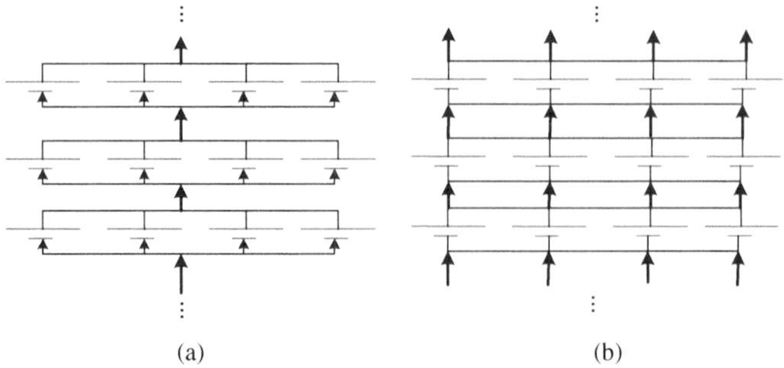

Fig. 6.7 Types of connections between modules of a battery pack: (A) single, (B) multiple.

In the PS configuration, when one cell arranged in parallel with other cells in a module becomes an open circuit, the rest of the cells in the module continue to operate, but the module capacity declines by an amount of one cell capacity, which reduces the capacity of the entire battery pack. If the failure in the module is not fixed, additional capacity fade will likely occur in the battery pack as the other cells in the modules are charged/discharged at higher rates, which accelerates the degradation rate within the battery cells.

Another failure that can occur in a PS configuration is that one of the cells in the module becomes a short circuit. In this case, the output voltage of the module becomes zero, which results in a considerable reduction in the capacity of the overall pack. At this condition, the drive-train unit connected with the battery pack can operate only at low load conditions, in which the required voltage is smaller than at high load conditions. However, the short circuit issue further decreases the capacity in the pack. Note that the repairing process which includes the replacement of the module is easier compared to the replacement of a single cell in the open circuit problem.

Finally, a failure in the serial connection between two modules results in operation failure of the entire battery pack. It may require replacement of rigid and flat conductors know as bus-bars, which carry current. A bus-bar is a solid metal bar and it is typically made from copper or aluminum.

To prevent failures of the types described in this section in the cells, modules, or packs in a battery system, the use of a battery management system is essential, as it monitors and provides data on the voltage, current, and temperature of the cells and balances the current/voltage distribution in minor failure conditions or stops operation of battery in severe conditions.

6.4 Mechanical design of battery pack

A suitable mechanical design for a battery pack provides a safe structure for extreme conditions, low cost and high productivity and reliability, ease of assembly and services. A suitable mechanical design usually results in a compact and lightweight battery pack. The pack should also be designed so as to allow for routing of cell vent gases to a safe exit and to contain possible cell failures. Furthermore, the pack design should allow for flow paths for the thermal management system (e.g., employing fin or cold-plate cooling).

Fig. 6.8 shows the main components of a battery pack. These components are now described:

- The *insulation pads* are placed between the modules connected in series to prevent leakage current (local current loop between subsequent modules). It is not suggested to use insulation pads between parallel cells in the modules because the insulation pads have both thermal and electrical conductivities. This means that, with the use of an insulation pad, the generated heat in the battery cells cannot be dissipated from the module and consequently from the battery pack.
- The *base plate* is responsible for holding the cells and modules together within the battery pack as well as protecting the cells from vibrations.

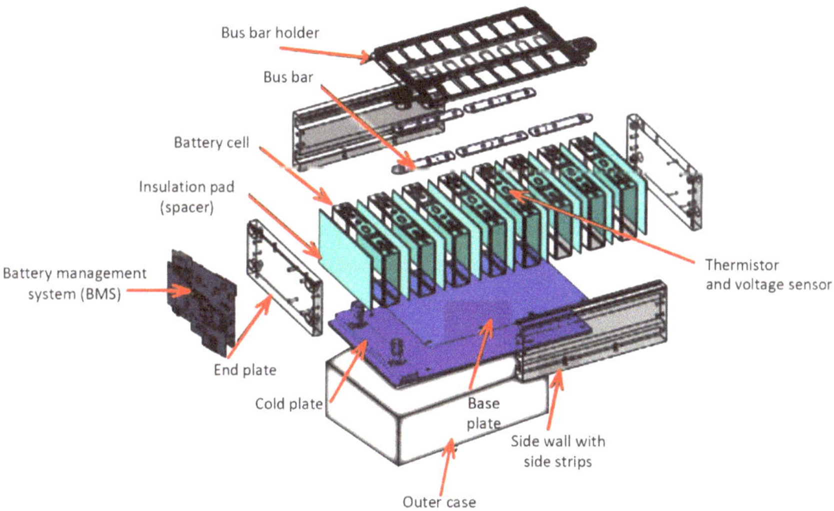

Fig. 6.8 Components of a battery pack. *(Adapted from Smith, J., Hinterberger, M., Hable, P. Koehler, J., 2014. Simulative method for determining the optimal operating conditions for a cooling plate for lithium-ion battery cell modules. J. Power Sources 267, 784–792.)*

- *Side strips* are used in the battery pack to help fix together the end plates. The basic function of side strips and end plates is to constrain movements of cells and to apply some pressure on the surfaces of the cells to prevent bulge phenomenon in the cells. Doing so can increase the cell lifetime.
- The *cold plate* is required in the battery pack to maintain the temperature of the cells within the desired operating range.
- Generally, the cells and modules are connected together by a *bus-bar*. The bus-bar is a metallic (copper or aluminum) strip or bar for current power distribution. In other words, bus-bars are needed for series and parallel connections of cells and modules to carry the current efficiently. The bus-bars should not move in the pack as that can lead to short circuiting in the battery. A *bus-bar holder* can be used to hold the bus-bars.
- A *temperature sensor (thermistor)* and *voltage sensor* are connected to the battery management system to transfer temperature and voltage data, based on which required actions are made by the battery management system.
- A *wire harness* is used for efficient flow of signals and energy. The harness secures wires against the effects of vibrations, moisture and abrasions, reducing the risk of an electrical short.
- A *battery management system* can be treated as the heart of the battery pack. The battery management system communicates with other electrical units in the pack like sensors for receiving information about temperature, voltage, and current flow distributions through the cells. The battery management system is also connected with units outside of battery pack like the motor and controller to order necessary actions.
- The *outer casing* is mainly used to protect the battery pack against the external environmental conditions and prevents water, moist, or dust from entering the battery pack. These environmental elements may increase the resistance within the bus-bar connections and cells, which can lead to a short circuit or higher generated heat inside the cells, and consequently reduce the lifetime of the pack. In addition, the outer casing protects the overall battery pack including modules, cells, and connections against sudden shocks like those that can occur during a crash of an electric vehicle.

The main important considerations for mechanical design are as follows:

(i) material selection,
(ii) base plate design for individual cell accommodation,
(iii) cell movement constraint and control,
(iv) uniform pressure over the surfaces of cells,
(v) outer case design for overall protection,

(vi) bus-bar design and
(vii) packaging constraints (e.g., available space for battery pack).

Generally, the criteria for material selection are cost effectiveness, broad availability, good strength and nonpoisonous. In the next section, the main forces acting on a battery pack, as well as the design procedures and criteria for a base plate, are comprehensively described.

6.4.1 Stress-strain theory

The main forces acting on the battery pack are shown in Fig. 6.9. They consist of tension forces on the side strips, bending forces on the end plate and base plate and shear forces acting on the welded joints. A tension force acts on the side strips to pull the end plates to create some pressure on the surfaces of cells. A bending force is associated with the battery pack weight on the base plate. When the bus-bars are welded on the cell terminals, any movement of the cells results in a shear force acting on welded joints.

Generally, there are five types of fundamental loadings, including compression, tension, shear, torsion, and bending. Fig. 6.10 shows these five types of forces acting on an element. Furthermore, *stress* is the internal resistance offered by the body of element to an external load (shown in Fig. 6.10) applied to it per unit cross section area. That is,

$$\sigma = \frac{F}{A} \tag{6.4}$$

where σ denotes stress, F force, and A original cross section area. The reason that the original cross section area is used is that the application of an external

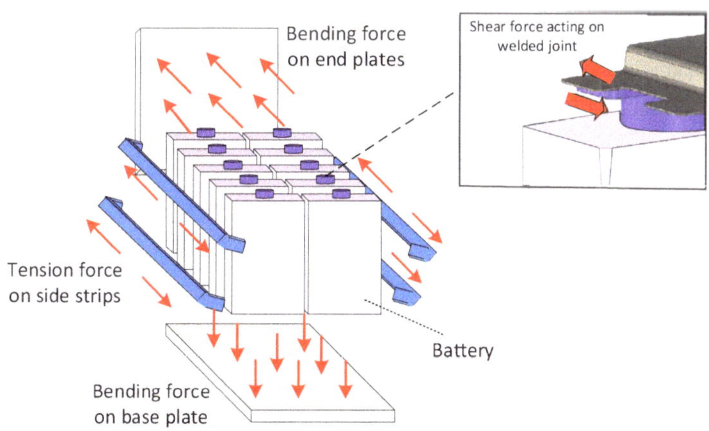

Fig. 6.9 Main forces acting on a battery pack.

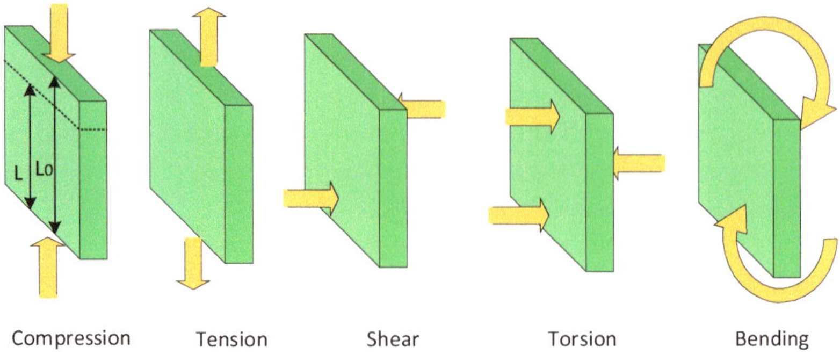

Fig. 6.10 Five types of forces acting on an element.

load to the element results in some elongation in the element which changes the cross-sectional area from its original form.

Due to the action of forces on an element, there would be some change in the length of element perpendicular to the cross-sectional area. This deformation of the element is the *strain* that results from the applied stress. It can be expressed as:

$$\varepsilon = \frac{L - L_0}{L_0} \tag{6.5}$$

where ε denotes strain, L length after load is applied, and L_0 original length. Terminologically, the strain expresses the elasticity of a module, and indicates how much an element can compressed, extended, twisted or bent, etc.

6.4.2 Design of base plate

The minimum requirement of the base plate is to tolerate the weight of battery cells and other components without cracking or undergoing extreme bending. If the extent of bend in a base plate is excessive, the base plate cannot constrain the battery cells. This leads to the movement of the cells and consequently damage of the cells and failure in operation of the bus-bar.

For determination of the base plate dimensions, it is often assumed that the weight of the battery pack is distributed uniformly. In addition, the weight of the batteries is the sole main force on the bus-bar neglecting the effects of vibrations (which are discussed in Section 6.4.3). With the arrangement of cells in the pack, the length and width of the base plate can be obtained, but the thickness of plate requires more extensive engineering calculations. The thickness of the base plate can be determined based on the load (weight of battery cells over total area of the base plate) and the selected material for the base plate.

There is a technique for selection of the material for a base plate which is known as the Ashby methodology (Ashby, 2017). This methodology is based on the objective functions which in this case are cost, strength, weight, and insulating properties. A procedure is defined in this methodology as shown in Fig. 6.11. There, the translation means the definition of objectives (i.e., strength, cost, and weight). In the screening step, the maximization or minimization of objectives (i.e., maximum strength, minimum cost, and weight) is defined. In this step, some of the materials can be screened and then removed from the list based on designer experience and other desirable factors. This helps to shorten the list of materials that can be used for construction of the base plate.

Alternatively, all the objective functions can be converted to maximization or minimization definitions, for example minimization of cost means maximization of 1/cost. In this case, the aim is to minimize the deflection of the plate which is proportional to modulus of elasticity, which is the ratio of stress to strain Y. Also, the modulus of elasticity is inversely proportion to the plate thickness t (i.e., $Y \propto 1/t^3$). Therefore, to minimize the deflection of the plate, the strain to the power of 1/3 should be maximized. So, the material index for the base plate can be defined based on the maximization of $Y^{1/3}$, and minimization of cost and density, which can be written as (Ashby, 2017):

$$\text{Material index for flat base plate: maximize } \frac{Y^{1/3}}{C \cdot \rho} \tag{6.6}$$

where E denotes modulus of elasticity (also known as Young's modulus), C cost of material, and ρ density of material. In the ranking step, the materials are ranked based on the maximum material index. A supporting information

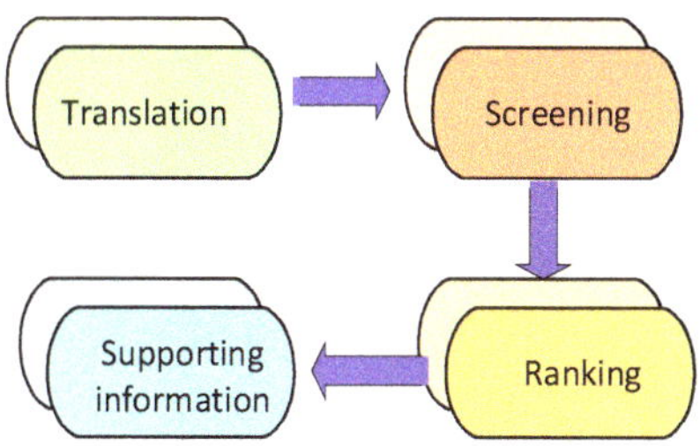

Fig. 6.11 Procedures in Ashby methodology for selection of material for the base plate of a battery pack.

step is used when two or more materials have about the same index; then other objectives (i.e., secondary objectives) like insulation properties can be considered in the material selection procedure to exclude undesirable materials. For example, when aluminum and acrylonitrile butadiene styrene (ABS) plastic have the same ranking (same material index), other objectives like capability to remove heat can be considered so the aluminum would be the preferred material as it has a higher thermal conductivity than ABS plastic. Table 6.2 compares two materials, mild steel and aluminum 6061, in terms of cost, density, and Young's modulus. It can be seen that the material index is higher for mild steel than aluminum 6061. So, the properties of the selected material (i.e., mild steel) are used for the plate's thickness calculation.

Based on the above methodology, the material can be selected for the base plate. The next step is to determine the thickness of the plate. According to Kirchhoff's equation in bending theory for plates, the maximum deflection and stress can be expressed as follows (Young and Budynas, 2020; Shanmugam and Wang, 2007):

$$y_{\max} = \frac{-\alpha q b^4}{E t^3} \tag{6.7}$$

$$\sigma_{\max} = \frac{-\beta q b^2}{t^2} \tag{6.8}$$

where q denotes load (in this case weight of cells over area of plate), b width of plate, and t plate thickness. The parameters α and β are determined based on the aspect ratio (ar) which is defined as the ratio of length a to width b (i.e., $ar = a/b$). The values of α and β can be found based on the design data available elsewhere (Young and Budynas, 2020; Shanmugam and

Table 6.2 Comparison of mild steel and aluminum 6061 in terms of cost, density, and Young's modulus.

	Material	
Parameter	**Mild steel**	**Aluminum 6061**
Density (kg/m^3)	7850	2700
Young's modulus (GPa)	210	68.9
Cost of material ($/kg)	0.32	3.69
Material index	21.8×10^{-3}	5.34×10^{-3}

Data in this table are drawn by the present authors from many articles and websites, and represent typical or average values.

Wang, 2007; Belkhodja et al., 2020) for different shapes and materials of the plate. In addition, the deflection can be a constraint based on the desired design conditions. For example, it can limit the maximum deflection in the downward direction to not more than a certain value (typically no more than 3 nm in many applications).

Next, the thickness of the base plate, as calculated with Eq. (6.7), can be used for determination of the maximum stress using Eq. (6.8). The maximum stress should not exceed the allowable bending stress for a given material. In the case that we want to decrease the base plate thickness, while maintaining the strength of the plate, several ribs or beams can be used to support the base plate.

In addition to the base plate design considerations, the appropriate design of the *end plate* is necessary for an optimum pressure distribution among the battery cells. Applying pressure on the surfaces of cells improves the cells lifetime as it prevents the cells from bulging. Typically, battery cell manufacturers provide an optimum pressure range that should be applied to the surface of a cell for long operation of a battery. For instance, for a lithium ion prismatic cell a small amount of external pressure, normally 10–20 kPa, is desirable, while cylindrical cells maintain pressure on the jelly roll (a cell assembly in which long coils of the electrodes are wound on a central core) without the need for external compression (Santhanagopalan et al., 2014). The pressure should be applied by the strips and end plate as shown in Fig. 6.12. Thus, the suitable thickness for the end plate as well as dimensions of strips are important for a uniform distribution of optimum pressure.

6.4.3 Vibration isolation

A battery pack should be designed so as to isolate transmission of undesirable vibrations to the cells, especially in moving applications of batteries

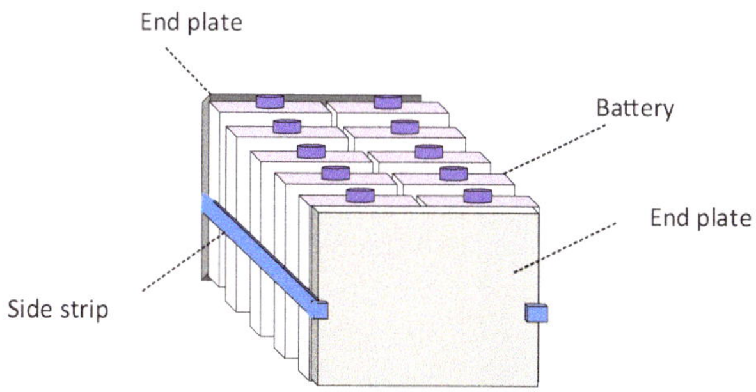

Fig. 6.12 Use of side strips and end plates to apply pressure on batteries in a pack.

(e.g., electric vehicles). The vibrations transmitted to the battery cells usually has a frequency equal to or less than 100 Hz (Watanabe et al., 2012). The resulting dynamic mechanical loads on the bus-bars and terminal connectors can lead to electrical continuity loss and fatigue failure of the casing. In addition, a dangerous problem can be caused when the natural frequency of the battery structure is near a frequency range of 100 Hz. This results in resonance that causes delamination of battery layers, shortening the battery lifetime. Several standard tests, as well as real-condition loads, are used to identify how vibrations and shocks affects battery cells in moving applications (Brand et al., 2015). Therefore, a proper vibration isolation of electronic subsystems is required to prevent a failure of the durability of battery packs in electric/hybrid electric vehicles.

When an electric/hybrid electric vehicle starts to move on the road, continuous vertical vibrations with low frequency are transmitted to the battery pack. In addition, electric/hybrid electric vehicle movements over holes, uneven road surfaces and bridge abutments cause shocks that results in mainly vertical vibrations as well as some lateral vibrations. So, vertical and lateral supports are required to stabilize the battery cells in the battery pack. To achieve this, pressure is applied on the top surface of the battery pack.

In addition to the outer case, an inner case is also often used in construction of a battery pack. The inner case can be made of a polymer material to provide an insulation layer around the battery pack. The inner case is a combination of top and bottom shells which are constrained by spacers placed between the battery cells. Typically countersunk head screws are used in these spacers to pull the top and bottom shells together (see Fig. 6.13).

To protect the inner case from vibrations and shocks, L-shape mounts can be installed on the edges of the inner case to absorb vibrations and provide damping (see Fig. 6.14). Then the whole package (mounts and inner case) are placed inside the outer case.

Another design, shown in Fig. 6.15, was proposed for supporting the battery cells within a battery pack in an electric vehicle by using a cover-pad-tray retention arrangement (Zhou et al., 2009). This frame consists of four beams which engage the four sides of the battery pack. Tensioning bolts are used to connect the frame with the battery pack. Two types of flat and L-shape damping pads are used to absorb vibrations along the vertical direction and prevent movement of the modules upward and downward. These pads are made of a nonslip material to provide sufficient friction force to eliminate movement. The L-shape damping pads are located at each of

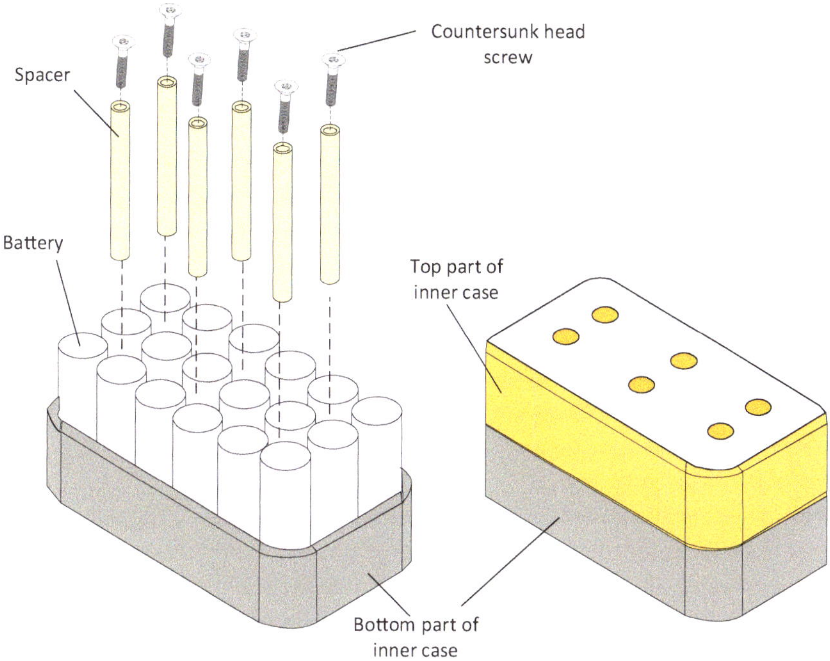

Fig. 6.13 Inner casing of a battery pack containing cylindrical battery cells. Countersunk head screws are used in the spacers to pull the top and bottom shells together.

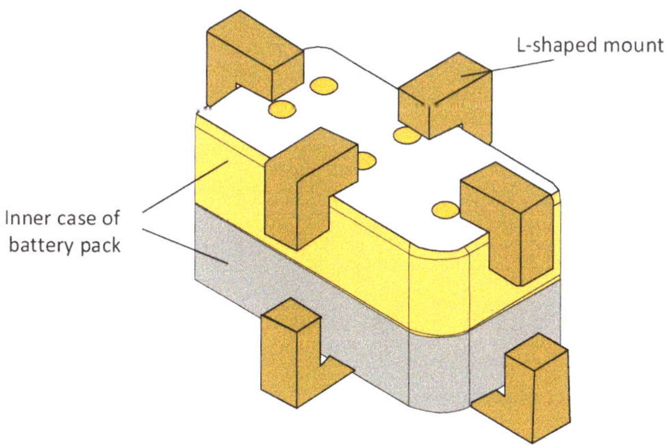

Fig. 6.14 L-shape mounts installed on the inner case to absorb vibrations and provide damping.

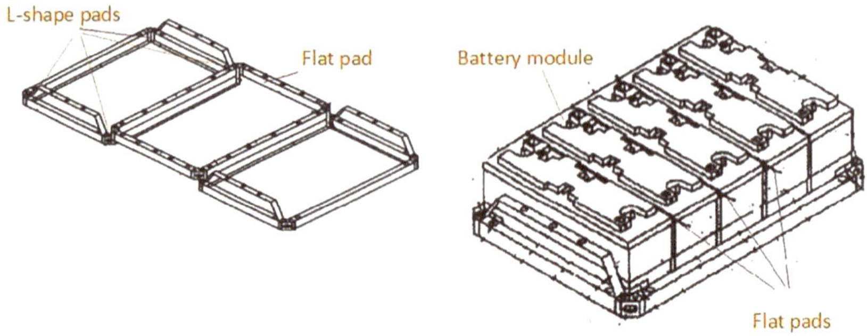

Fig. 6.15 View of a supporting frame arrangement used with a battery pack: (left) an empty frame, (right) the battery modules placed in the frame. *(Modified from Zhou, S., Husted, C.C., Benjamin, F.A., 2009. Battery Pack Arrangements. U.S. Patent 7,507,499.)*

the corner connections and bear against the frame structure to apply a small pressure at each of the corners and push the battery modules together. The flat shape damping pads are placed between individual modules, as well as at the upper and lower corners of facing sides of adjacent battery modules. The tensioning bolts are fastened after assembling the frame to compress the modules toward one another and squeeze the pads positioned between the modules to prevent any movements.

A mounting frame can be also used in the battery pack for uniform distribution of vehicle weight and for maintaining a low center of gravity. Fig. 6.16 shows the design of a typical mounting frame (Iwasa et al., 2011). In the two front rectangular areas of the mounting frame, the battery cells are arranged in vertical directions so that the long side is placed in the transverse direction of the electric vehicle and the short side is placed in the longitudinal direction. In the rear area of mounting frame, the battery cells are arranged in such a way that the shortest side is placed in the transverse direction of the electric vehicle. The weight of battery cells arranged in the rear area of mounting frame is equal to the weight of cells placed in the front area of the frame. At this condition, the center of gravity of the battery pack is located at the rear of the electric vehicle, which is desirable in terms of weight balancing when the electric motor, inverter, and battery charger are placed in the front of the compartment. This mounting frame can be used for different types of vehicles in which the layout of seats is not the same. By identifying the optimum weight distribution, the vibration isolation can be modified by changing the number of battery modules without necessitating major changes in the dimensions of the mounting frame.

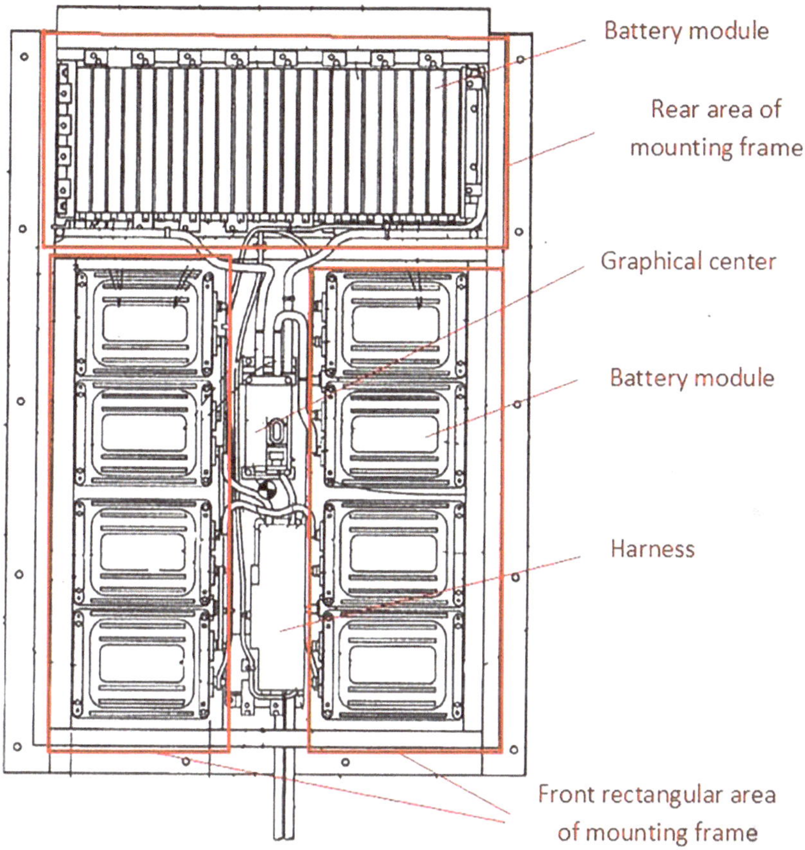

Fig. 6.16 View of a battery assembly in a mounting frame. *(Modified from Iwasa, M., Kadota, H., Hashimura, T., Shigematsu, S., 2011. Vehicle Battery Mounting Structure. U.S. Patent, 300426.)*

6.5 Thermal design of battery pack

The thermal design of battery cells in a pack is important for maintaining the battery cells within their safe and efficient operation temperature range. Thermal runaway is one of the main concerns for safe operation of battery cells, especially in moving applications like electric/hybrid electric vehicles. Thermal runaway can be caused by impurities in the chemical compounds present in the construction of battery cells, improper charging and discharging (e.g., fast charging or discharge below the limit specified) or ineffective battery thermal management. One of the indicators of thermal runaway is a rapid temperature increase of the battery cells. Another indicator is gas leakage from the battery pack; if a proper ventilation unit has been installed in the pack, this gas leakage can be vented from the pack. A rapid increase of battery temperature beyond its optimum value directly affects the electrochemical reactions occurring in the cells, which reduces the battery's efficiency, the amount of charge the battery cells can accept, and the power and available capacity of the batteries. All of these factors ultimately shorten the lifetimes of the battery cells. In addition, in the condition of thermal runaway, the battery cell and pack materials may lose their strength and some internal components like bus-bars may melt. These undesired consequences emphasize the importance of using an effective and efficient battery thermal design.

The thermal design of a battery pack should lead to a thermal management system that is compact, lightweight, easily packaged and serviced, low in energy consumption and low cost. An effective thermal design of battery pack maintains the optimum temperature of the battery cells and consequently the battery pack, which results in an increase in the lifetime of the battery cells.

The other main aim of the thermal design of a battery pack is to provide a *uniform temperature distribution* between the battery cells, so as to eliminate local hot spots in the pack. The more uniform the temperature distribution throughout the battery pack, the longer is the lifetime of the battery cells. The thermal design of a battery pack should be carried out in such a way that all battery cells (at the edges, middle, or corners) have sufficient heat transfer surface exposed to the cooling medium through which the generated heat inside the cells can be removed effectively.

Fig. 6.17 shows the impacts of temperature on lithium ion battery cells in an electric vehicle. It is seen that relatively higher temperatures degrade the lithium ion battery cells more quickly. So, cooling of the battery cells is necessary for continuous operation of battery cells, especially during fast charging, hot weather, and moderate to large current demands. Also, low operating temperatures reduce the power and energy capabilities since the internal resistances of the battery cells are relatively high. So, heating of battery cells is required for charging and discharging of batteries during low operating temperature conditions, in order to maintain the battery cell temperatures within the desired range (e.g., between 15°C and 35°C for the lithium ion battery).

For the thermal design of a battery pack, one should identify how much heat is generated inside the battery pack. Heat is mainly generated within the battery cells due to (i) resistance to electrochemical reactions (i.e., heat

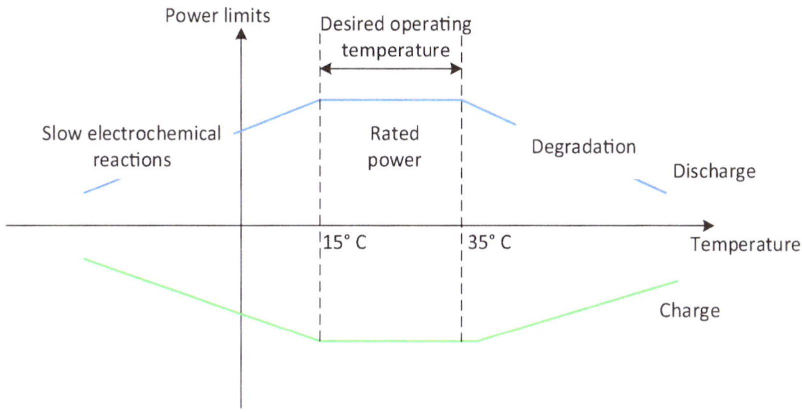

Fig. 6.17 Lithium ion battery delivered/input electrical power profile with temperature of battery cell in an electric vehicle. *(Modified from Pesaran, A., 2016. Battery pack thermal design. In: Presentation at 16th Annual Advanced Automotive Battery Conference, Detroit, US. Available from: https://www.osti.gov/servlets/purl/1304580 (Accessed 26 May 2022).)*

generation during charge transfer between electrode and electrolyte in electrode reactions), (ii) resistance to flow of electrons (i.e., Joule or ohmic heating), and (iii) resistance to movements of ions in the electrolyte. Heat generation in battery cells is extensively described in Chapters 2–4. Heat is also generated in other elements placed in the battery pack, such as bus-bars and a battery management system, primarily due to internal resistances in such elements (i.e., Joule heating). The Joule heating rate in any element, with an internal electronic resistance R when a current I is circuited through this element, can be expressed as:

$$\dot{Q}_J = I^2 R \tag{6.9}$$

From Eq. (6.9), it can be inferred that the Joule heating rate in an element depends on the rate of charge/discharge, as well as the internal resistance of the element. The internal resistance for an individual battery cell depends on the operating temperature, depth of discharge, and open circuit voltage. For the bus-bar and battery management system (including power MOSFET, contactors, etc.), the internal resistance primarily depends on the operating temperature. By determination of the overall heat generated within a battery pack, how much heat load should be removed or added from/to the pack can be found. Next, an example is provided for heat load determination in a battery pack.

Table 6.3 Prismatic battery cell characteristics.

Parameter	Value
Nominal capacity	15 Ah
Nominal voltage	3.65 V
Internal resistance	10 mΩ

6.5.1 Example: Heat load determination in a battery pack

Consider a battery pack containing two parallel rows each row consists of 16 prismatic battery cells connected in series (i.e., 2P16S). The characteristics of these prismatic battery cells are shown in Table 6.3. If the pack is discharged from 100% to 20% (state of charge, SOC) at 1C (discharge rate), calculate the heat generated and energy expelled by the pack. Also, compare the energy lost as heat to total pack energy during discharge.

Solution:

For determination of the heat generation rate within the battery cells, it is assumed that most of the generated heat is associated with Joule heating and that other contributors can be neglected. So, the total heat generation rate within the battery cells in the pack can be written as:

$$\dot{Q}_{J,T} = N_B I^2 R \tag{6.10}$$

where N_B denotes total number of battery cells (i.e., $N_B = 32$), I current drawn from the cells during discharge, and R internal resistance of the battery cell. A discharging current of 1C can completely discharge a fully charged battery in 1 h. So, the current that is drawn from a battery cell at 1C discharge would be 15 A. Substituting the values of current, internal resistance and number of cells in Eq. (6.6) yields total heat generation rate in the battery cells in the pack. That is,

$$\dot{Q}_{J,T} = 32 \times 15^2 \mathrm{A}^2 \times \left(10 \times 10^{-3}\right)\Omega = 72 \text{ W} \tag{6.11}$$

The amount of heat that is lost by the battery pack can be determined as:

$$Q_{J,T} = \dot{Q}_{J,T} \times t \tag{6.12}$$

It can be assumed that there is a linear relationship between time and discharge rate. This means that, when the battery cell is fully discharged (100%–0%) in 1 h (3600 s) at 1C, it would be discharge from 100% to 20% in 0.8 h (2880 s). Thus, the amount of heat that is generated during 2880 s can be determined:

$$Q_{J,T} = 72 \text{ W} \times 2880 \text{ s} = 207.36 \times 10^3 \text{ J} = 207.4 \text{ kJ} \tag{6.13}$$

By comparing this amount of heat generation in the battery cells with total energy (both heat and electrical energies) that are produced by the battery pack (i.e., $N_B \times V \times I \times t = 32 \times 3.65\ \text{V} \times 15\ \text{A} \times 2880\ \text{s} = 5045.8\ \text{kJ}$), it can be observed that 4.1% of total energy is lost as heat in the battery cells (i.e., $(207.4\ \text{kJ}/5045.8\ \text{kJ}) \times 100\% = 4.1\%$). Therefore, it is found that 4.1% of total energy of the battery pack is lost as heat due to internal resistance within the battery cells during discharge from 100% to 20% SOC at 1C. If this amount of heat is not removed from the pack, it likely will lead to a rapid increase in the operating temperature of battery pack. Note that the heat generation rate in the battery cells can be reduced by lowering the discharging rate (i.e., 0.1C) or selecting other battery cells with better chemistries in terms of lower internal resistance.

6.5.2 Battery thermal management systems

The thermal design of a battery pack includes the design of an effective and efficient battery thermal management system. The battery thermal management system is responsible for providing effective cooling or heating to battery cells, as well as other elements in the pack, to maintain the operating temperature within the desired range, i.e., the temperature range at which the battery pack operation is safe and efficient. An optimum design of a battery thermal management system can potentially extend the lifetime of the battery pack. Battery thermal management systems are generally divided into two categories: active and passive. The active mode of thermal management includes (i) air-based, (ii) liquid-based, and (iii) refrigerant-based systems or combination thereof. Passive thermal management systems typically include phase change material-based systems, which can be based on liquid-to-vapor or solid-to-liquid phase change processes. Fig. 6.18 shows the main types of battery thermal management systems. Various battery thermal management systems and their working principles, applications, advantages, and drawbacks are comprehensively described in Chapter 5.

Most *air-based* battery thermal management systems are designed based on forced convection air flow, so that is the focus of this section. In an air-based battery thermal management system, a fan or blower is typically used to circulate air around the battery cells then to reject it to the environment. These systems are low in cost and have simple configurations with easy maintenance. Typically, air-based battery thermal management systems are capable of providing cooling for moderate cooling load applications.

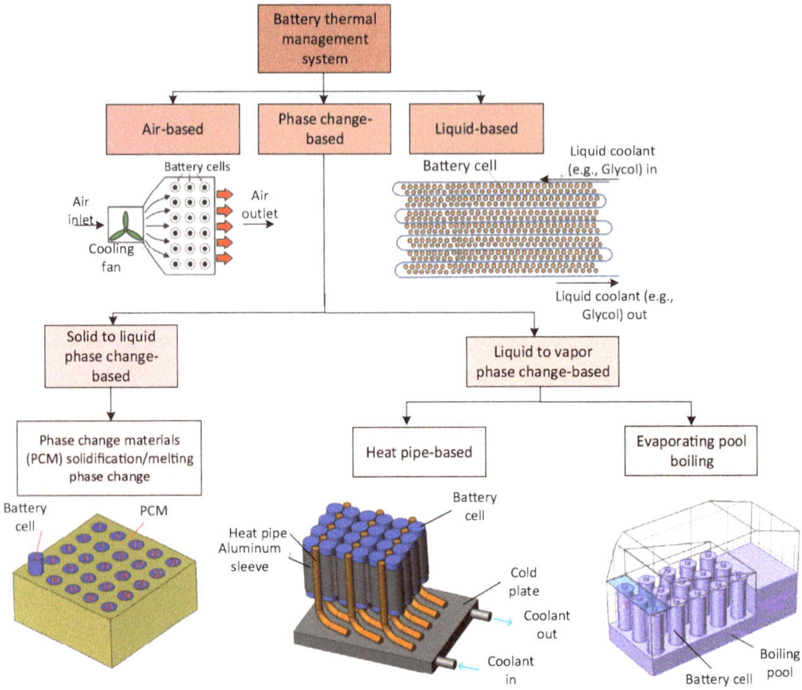

Fig. 6.18 Main categories of battery thermal management systems.

Liquid-based battery thermal management systems are capable of removing relatively high heat loads, especially at high C-rate operating conditions, compared to the air-based systems. This is because the working fluid is a liquid (usually water or an ethylene glycol/water mixture), which has a higher convective heat transfer coefficient than air (i.e., $20\,W/m^2\,K < h_{air} < 100\,W/m^2\,K$ and $300\,W/m^2\,K < h_{water} < 3000\,W/m^2\,K$). In liquid-based battery thermal management systems, a chiller is required to cool water, which requires the use of a significant amount of energy. Liquid-based cooling systems are the most commonly used battery thermal management systems for electric and hybrid electric vehicles.

PCM-based battery thermal management systems include systems based on solid-liquid phase change and liquid-vapor phase change. Conventionally, the PCM is more often used for solid-liquid phase change, so this chapter focuses on that type. Thus, the PCM is taken in this chapter to be a substance that absorbs and releases energy by melting and solidifying, respectively, at the phase change temperature. In liquid-vapor phase change-based systems, heat pipes can be used to convey heat from the battery to a condenser via a heat transfer fluid, or batteries can be directly submerged in a stationary liquid heat transfer fluid. The former is known as a heat pipe-based battery thermal management system and the latter is known as an evaporating pool boiling-based battery thermal management system.

In addition, *refrigerant-based* battery thermal management systems constitute a type of PCM-based battery thermal management system that is capable of removing high heat loads at high C-rate operating conditions compared to air-based and liquid-based battery thermal management systems. In a refrigerant-based system, which consumes electrical energy to provide the cooling effect, the refrigerant undergoes a phase change process in which it absorbs the high heat loads generated in battery cells. When an environmentally benign and inexpensive refrigerant is used and leakage in the refrigeration cycle is controlled, this system can provide an efficient thermal management system in electric vehicles.

The *PCM-based* battery thermal management system has relatively high energy density, and can absorb a high amount of heat from the battery cells relative to other thermal options, on an equivalent mass basis. As well, a PCM-based battery thermal management system can provide a uniform temperature distribution throughout the battery pack. Such thermal management systems can be considered as passive, in that they can store and/or release large amounts of thermal energy with no additional energy consumption. However, the solid-to-liquid phase change material used in such systems (like paraffin) has a low thermal conductivity, and it expands when it absorbs heat. Thus, some metal bars can be used in PCM-based battery thermal management systems to enhance the thermal conductivity of the system. Also, extra volume should be allocated for the PCM to account for expansion. A concern in a PCM-based battery thermal management system is the melting temperature of the PCM. When a particular PCM has a melting temperature below the ambient temperature, it absorbs heat from the ambient, melts and loses its ability to function for cooling the battery cells.

Other types of battery thermal management systems have been developed based on selective integrations of the above-listed three main categories and these are known as *hybrid or integrated* battery thermal management systems. For instance, a PCM-based system can be integrated with a liquid-based or air-based system. In this combined system, the PCM-based system absorbs heat from a set of batteries, and is then cooled by a liquid-based or air-based thermal management system. Often, a uniform temperature distribution can be achieved with this integrated thermal management system through the provision of uniform cooling to the battery and maintaining the temperature of battery at the desired operating condition.

In all mobile applications of battery systems, including marine, aviation and road vehicles, thermal management of battery cells is an important factor in vehicle design. The battery thermal management system maintains the battery temperature within the desired operating range. There has been much research on battery thermal management systems. For example, several studies have been reported recently for systems that are air-based (Zhang et al., 2021; Lin et al., 2021), liquid-based (Kalaf et al., 2021; Rao et al., 2017), solid-liquid phase change based (Chen et al., 2021; Malik et al., 2016), and liquid-vapor phase change based (Al-Zareer et al., 2019; Wang et al., 2016). A useful review on novel battery thermal management system has been reported (Al-Zareer et al., 2018).

6.5.3 Selection of battery thermal management system

In the previous section, various types of battery thermal management systems were presented. Now, it is important to determine which types of these systems are suitable for use in particular applications. There are many criteria for selection of a suitable battery thermal management system for a given application. For an electric vehicle, for instance, some factors like the heat generation in the battery cells in the pack, compactness, weight, effective cooling/heating effects, and cost are important criteria for selection of an appropriate battery thermal management system. Among these factors, the heat generation rate in the battery pack is a determining element when choosing a thermal management system in an electric vehicle. The battery thermal management system should be able to maintain the battery cells within the safe and efficient temperature range. In the remainder of this section, a method for selecting a battery thermal management system based on the heat generation rate in the battery pack is described.

The main step in determining a suitable battery thermal management system based on the heat generation rate inside the battery cells is to evaluate the heat transfer coefficient that is required for an effective convective heat transfer process between the cells and the heat transfer medium. This can be determined based on the thermal resistance and equivalent thermal circuits. The thermal resistance concept states that, when heat flows between two points with different temperatures, there is a resistance to the heat flow between these two points.

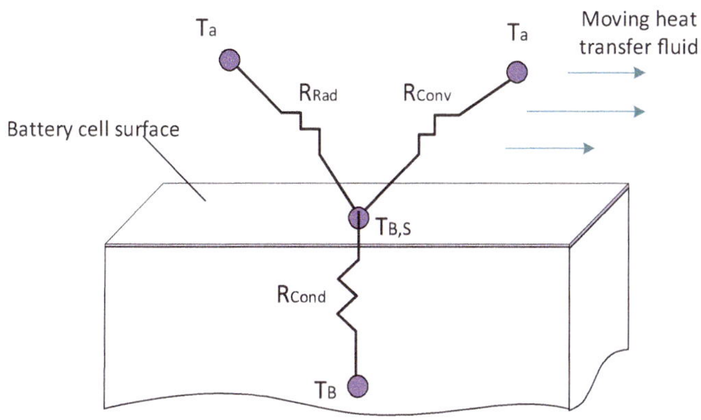

Fig. 6.19 Electrical circuit analogy for a battery cell in contact with a heat transfer fluid.

The heat flow can be modeled by analogy to an electrical circuit where heat flow is represented by current, temperatures are represented by voltages, heat sources are represented by constant current sources, thermal resistances are represented by resistors and thermal capacitances by capacitors. Fig. 6.19 compares the heat flow and electric flow approaches. The general thermal resistance in an element can be written in terms of heat transfer rate $\dot{Q}$ and temperature difference between two points ΔT. That is,

$$R = \frac{\Delta T}{\dot{Q}} \tag{6.14}$$

Generally, a thermal resistance can be observed in all three modes of heat transfer: (i) conduction, (ii) convection, and (iii) radiation. For instance, when a battery cell is in contact with a heat transfer fluid for cooling or heating purposes (see Fig. 6.20), the equivalent thermal resistance for these heat transfer modes respectively can be expressed as follows:

$$R_{\text{cond}} = \frac{L}{kA} \tag{6.15}$$

$$R_{\text{conv}} = \frac{1}{h_{\text{conv}} A} \tag{6.16}$$

$$R_{\text{Rad}} = \frac{1}{h_{\text{Rad}} A} \tag{6.17}$$

where L is the thickness of element through which conduction heat transfer occurs, k is the thermal conductivity, and A is battery area. Also, h_{conv} and

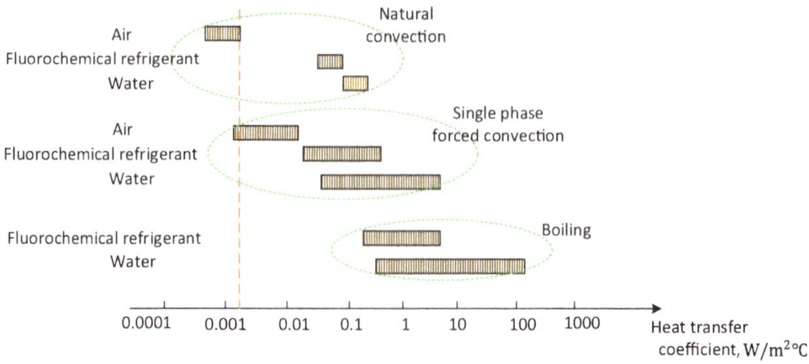

Fig. 6.20 Heat transfer coefficient ranges for various types of convective heat transfer. Data are drawn from many articles and websites, and represent typical ranges for various types of convective heat transfer.

h_{Rad} are the convective and radiative heat transfer coefficients respectively, which can be written as:

$$h_{\text{conv}} = \frac{q_{\text{conv}}}{(T - T_a)} \tag{6.18}$$

$$h_{\text{Rad}} = \frac{q_{\text{Rad}}}{(T - T_a)} = \varepsilon\sigma\left(T^2 + T_a^2\right)(T + T_a) \tag{6.19}$$

Here, q_{conv} and q_{Rad} respectively are the convection and radiation heat fluxes, while T_a denotes ambient temperature, T surface temperature, σ Stefan-Boltzmann constant, and ε surface emissivity.

Ideally, it can be assumed that the temperature distribution inside the battery cell is uniform, meaning that the temperature throughout the battery cell is constant (at T_B). So there would be no conduction heat in the cell. Also, radiation heat is insignificant compared to convective heat transfer. Then, based on the energy balance for the battery cell, all generated heat inside the battery is transferred to the heat transfer fluid through convection, i.e.,

$$\dot{Q}_{\text{gen}} = \dot{Q}_{\text{conv}} \tag{6.20}$$

Based on Eqs. (6.14) and (6.16), the thermal resistance associated with convection can be expressed as:

$$R_{\text{conv}} = \frac{1}{h_{\text{conv}}A} = \frac{T_B - T_a}{\dot{Q}_{\text{conv}}} = \frac{T_B - T_a}{\dot{Q}_{\text{gen}}} \tag{6.21}$$

By determining the heat generation rate in the battery cell (discussed in Section 6.4.2), the heat transfer coefficient can be determined with Eq. (6.17). Various graphical and other presentations of the relations between heat transfer coefficient and convection heat transfer methods (i.e., forced/natural air, water circulation, or boiling heat transfer) have been developed, based on experimental studies at certain design conditions.

Fig. 6.20 shows a typical graph that provides information and suggestions for the use of convective heat transfer for various ranges of heat transfer coefficient. For instance, recall the example in Section 6.5.1. The heat generation rate in the battery pack was found to be 72 W. For maintaining the battery cells at a temperature of 45°C, considering an ambient temperature of 35°C, the heat transfer coefficient for heat transfer between a battery cell with convective surface area of $0.168\,m^2$ would be 42.85 $W/m^2\,°C$ ($=72W/((45\,°C-35\,°C)\times 0.168\ m^2)$). When the heat transfer coefficient is considered as the only primary criterion for selection of a battery thermal management system, based on the graph in Fig. 6.20, the suitable thermal management method can be forced air convection.

6.6 Design integration and summary

Fig. 6.21 summarizes the stages of battery pack design which are described in detail in this chapter. The cell arrangements in the battery pack follow the PS (parallel-series) and SP (series-parallel) configurations. The required equipment for thermal, mechanical, and electrical designs are presented under the structure tab in Fig. 6.21. In addition, the safety considerations in each of the designs are located in the safety stage of the pack design. The control unit responsible for checking and monitoring the pack characteristics are part of the battery management system. The final stage is associated with the application, which includes the interfaces between different subsystems and monitoring and telemetry services, which allow all battery operating parameters to be monitored in real time. Telemetry is used in many battery operated wireless systems to inform monitoring personnel when the battery power is reaching a low point and the battery is nearing its end of life. Telemetry provides data relating to battery packs in real time (e.g., SOC, depth of charge, or the amount of energy that has been discharged) by monitoring the condition of the battery and ensuring it is operational for the entire vehicle service life.

6.7 Closing remarks

This chapter describes the main process steps for design of a battery system, including design of battery the management system, as well as electrical, mechanical, and thermal designs along with safety and cost considerations. In addition to prudent selection of materials used in the battery cells, an

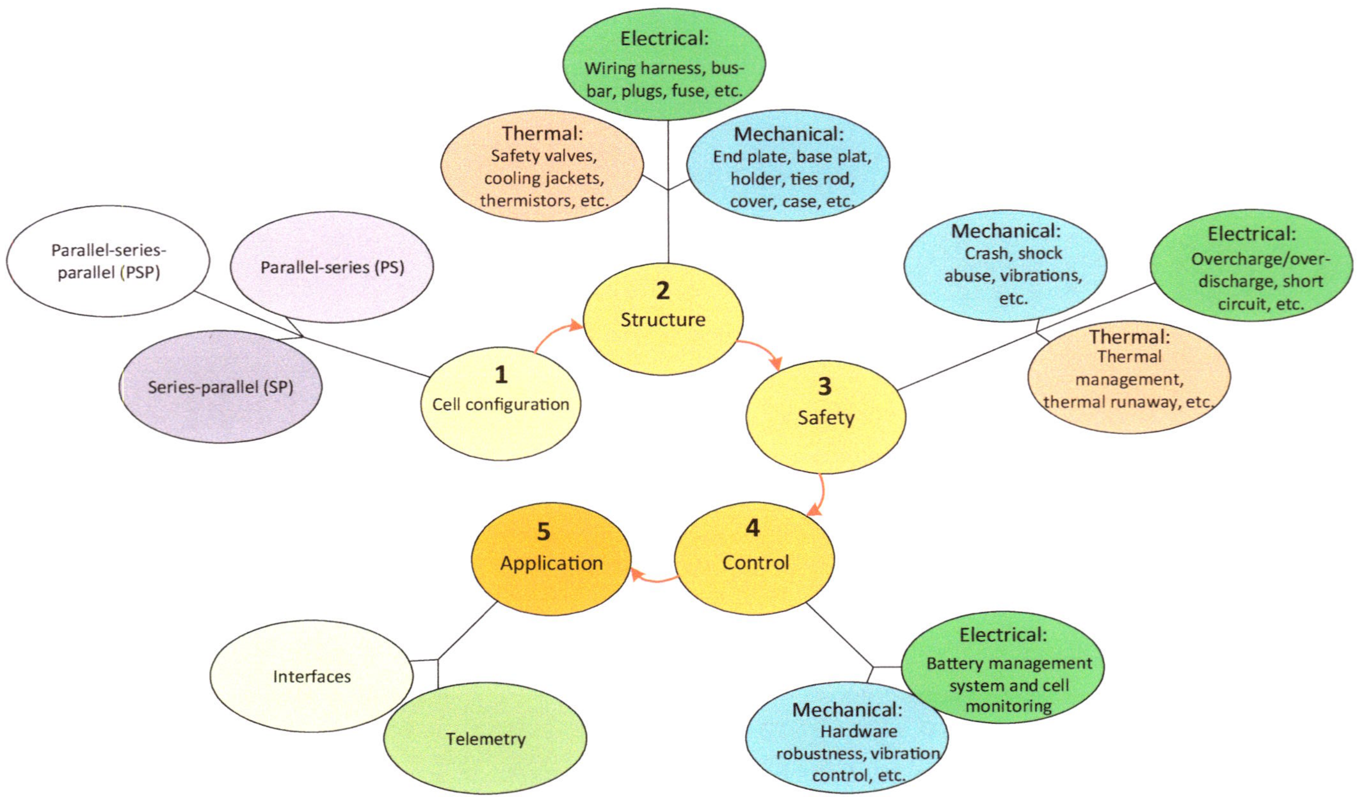

Fig. 6.21 Stages of battery pack design.

optimum design of a battery pack extends the lifetime of the battery cells and provides safe and reliable operation of the battery pack since doing so reduces uneven current distribution, poor control of discharge/charge and risks of failures.

The battery management system is responsible for monitoring the current, voltage, and temperature of batteries during the operation of system and halting the operation of the battery system when any of these operating parameters exceed from their limiting values. Also, this system uses current, voltage, and temperature data for determination of the state of charge and state of health of the battery pack. The proper electrical design of the battery pack yields an accurate evaluation of the electrical power that the battery needs to provide for a given application, as well as the balanced current/voltage distribution between the battery cells. A suitable mechanical design for a battery pack provides a safe structure for extreme conditions (e.g., vibrations, shocks) with minimum cost and weight and maximum reliability and compactness. One of the primary aims of mechanical design of a battery pack is to provide an optimum pressure on the surfaces of battery cells to avoid bulge in the cells while protecting them from vibrations and addressing other safety considerations. The effective thermal design of a battery pack leads to a uniform temperature distribution throughout the battery cells, and maintains the temperature within an optimum range by effectively removing the heat generated inside the battery cells.

Study questions

6.1. What are the primary design processes for a battery pack system?

6.2. What are the core functions of a battery management system?

6.3. What is the main function of power MOSFET in a battery pack design?

6.4. What are two main arrangements of battery cells in a pack? Describe the advantages and disadvantages of each of these configurations.

6.5. What are the main considerations for mechanical design of a battery pack?

6.6. What criteria should be considered while designing a battery thermal management system?

6.7. Compare refrigerant-based and liquid-based battery thermal management systems.

6.8. How is a suitable thermal management system for a battery pack selected?

6.9. Consider the example in Section 6.5.1, and calculate the heat generated and energy expelled by the pack when it contains 28 battery cells and each battery cell has an internal resistance of 12 mΩ.

References

Al-Zareer, M., Dincer, I., Rosen, M.A., 2019. Comparative assessment of new liquid-to-vapor type battery cooling systems. Energy 188, 116010.

Al-Zareer, M., Dincer, I., Rosen, M.A., 2018. A review of novel thermal management systems for batteries. Int. J. Energy Res. 42 (10), 3182–3205.

Ashby, M.F., 2017. Materials Selection in Mechanical Design, fifth ed. Butterworth-Heinemann, Amsterdam.

Belkhodja, Y., Ouinas, D., Zaoui, F.Z., Fekirini, H., 2020. An exponential-trigonometric higher order shear deformation theory (HSDT) for bending, free vibration, and buckling analysis of functionally graded materials (FGMs) plates. Adv. Compos. Lett. 29, 1–19.

Brand, M.J., Schuster, S.F., Bach, T., Fleder, E., Stelz, M., Gläser, S., Müller, J., Sextl, G., Jossen, A., 2015. Effects of vibrations and shocks on lithium-ion cells. J. Power Sources 288, 62–69.

Chen, K., Hou, J., Song, M., Wang, S., Wu, W., Zhang, Y., 2021. Design of battery thermal management system based on phase change material and heat pipe. Appl. Therm. Eng. 188, 116665.

Iwasa, M., Kadota, H., Hashimura, T., Shigematsu, S., 2011. Vehicle Battery Mounting Structure. U.S. Patent, 300426.

Kalaf, O., Solyali, D., Asmael, M., Zeeshan, Q., Safaei, B., Askir, A., 2021. Experimental and simulation study of liquid coolant battery thermal management system for electric vehicles: a review. Int. J. Energy Res. 45 (5), 6495–6517.

Lin, J., Liu, X., Li, S., Zhang, C., Yang, S., 2021. A review on recent progress, challenges and perspective of battery thermal management system. Int. J. Heat Mass Transf. 167, 120834.

Malik, M., Dincer, I., Rosen, M.A., 2016. Review on use of phase change materials in battery thermal management for electric and hybrid electric vehicles. Int. J. Energy Res. 40 (8), 1011–1031.

Rao, Z., Qian, Z., Kuang, Y., Li, Y., 2017. Thermal performance of liquid cooling based thermal management system for cylindrical lithium-ion battery module with variable contact surface. Appl. Therm. Eng. 123, 1514–1522.

Santhanagopalan, S., Smith, K., Neubauer, J., Kim, G.H., Pesaran, A., Keyser, M., 2014. Design and Analysis of Large Lithium-Ion Battery Systems. Artech House, London.

Shanmugam, N.E., Wang, C.M. (Eds.), 2007. Analysis and Design of Plated Structures: Dynamics. Elsevier, England.

Wang, Q., Rao, Z., Huo, Y., Wang, S., 2016. Thermal performance of phase change material/oscillating heat pipe-based battery thermal management system. Int. J. Therm. Sci. 102, 9–16.

Watanabe, K., Abe, T., Saito, T., Shimamura, O., Hosaka, K., 2012. Battery Structure, Assembled Battery, and Vehicle Mounting These Thereon. U.S. Patent, 8124276.

Young, W.C., Budynas, B., 2020. Roark's Formulas for Stress and Strain, ninth ed. McGraw-Hill Education, New York.

Zhang, F., Lin, A., Wang, P., Liu, P., 2021. Optimization design of a parallel air-cooled battery thermal management system with spoilers. Appl. Therm. Eng. 182, 116062.

Zhou, S., Husted, C.C., Benjamin, F.A., 2009. Battery Pack Arrangements. U.S. Patent 7,507,499.

CHAPTER 7

Integrated battery-based systems

Chapter objectives

- Illustrate and describe the integration of various battery-based technologies.
- Discuss the performance of integrated battery-based systems and demonstrate how it is affected by varying operating conditions.
- Present comprehensive modeling, analyses, and assessments of two case studies for integrated battery-based systems.
- Describe and discuss the overall performances of the systems covered in the case studies.
- Provide recommendations for further improvements of integrated battery-based systems.

Nomenclature

A	cell area (m^2)
a_i	activity factor of species i in the electrochemical reaction
c_p	specific heat at constant pressure (J/kg K)
c	salt concentration (mol/dm^3)
D	diffusion coefficient (m^2/s)
c_p	specific heat at constant pressure (J/(kg K))
E	electrical potential (V)
E_{OCP}	open circuit potential (V)
F	Faraday constant (C/mol)
$\bar{g}$	molar Gibbs free energy (kJ/kmol)
ΔG_C	activation free energy (kJ/mol)
g	gravitational acceleration (m/s^2)
$\bar{h}$	molar enthalpy (J/mol)
i	current density (A/m^2)
i_0	exchange current density (A/m^2)
i_L	limiting current density (A/cm^2)
I	current (A)
J_n	pore wall flux of lithium ions ($mol/(cm^2\ s)$)
k	constants for the reactions occurring at electrode
K_s	equilibrium constant for the reaction
k_r	thermal conductivity in radial direction (W/m K)
k_z	thermal conductivity in axial direction (W/m K)
L_a	base temperature lapse rate per kilometer of geopotential altitude (K/km)
$\overline{LHV}$	low heating value (kJ/kmol)
m	mass (kg)

Battery Technology
https://doi.org/10.1016/B978-0-443-18862-6.00008-2

M	Mach number
n_e	number of electrons in the balanced electrochemical half reaction
N	number of cells
$\dot{N}$	molar flow rate (mol/s)
P	pressure (Pa)
P_i	partial pressure of component i (Pa)
Pd	power density (W/m^2)
$\dot{Q}$	heat transfer rate (W)
$\dot{Q}_{gen}$	heat generation rate per unit volume for the battery (W/m^3)
r	molar conversion rate
R	universal gas constant (J/(mol K))
R_a	characteristic gas constant for air (J/(mol K)))
R_i	resistance of component i (Ω)
R_s	particle radius (m)
$\bar{s}$	molar entropy (kJ/kmol K)
ΔS	entropy change due to electrochemical reactions (J/K)
t	time (s)
T	temperature (K)
U_{O_2}	air utilization factor
U_{H_2}	hydrogen utilization factor
V	voltage (V)
V_B	volume of battery (m^3)
$\dot{W}$	work rate (W)
x	stoichiometric coefficient
x_i	molar concentration of species i
x_{sc}	steam to carbon ratio
y	stoichiometric coefficient
Z	altitude (m)

Greek letters

ϵ	volume fraction of polymer
ρ_B	density of battery (kg/m^3)
ρ_{int}	electrical resistivity of interconnect (Ω cm)
ρ_c	electrical resistivity of cathode (Ω cm)
ρ_e	electrical resistivity of electrolyte (Ω cm)
ρ_a	electrical resistivity of anode (Ω cm)
λ	effective water content of the membrane
α	transfer coefficient
δ	thickness (m)
σ	electronic conductivity (S/cm)
$\gamma_{\pm}$	activity coefficient
γ_M	electrode surface roughness factor
γ	Bruggeman coefficient
γ_a	specific heat ratio of air
κ_0	ionic conductivity of electrolyte (S/cm)
κ^{eff}	effective electrolyte ionic conductivity (S/cm)

∇t_+^0	transference number
Φ	electrochemical potential (V)
η	overpotential (V)
$\eta_{isen,\ T}$	isentropic efficiency of the gas turbine
η_{inv}	inverter efficiency

Subscripts

a	anode, activation, air
B	battery
c	cathode, concentration
cons	consumed
SOFC	solid oxide fuel cell
gen	generation
m	membrane
ohm	ohmic
ref.	reference
s	surface
+	positive electrode
−	negative electrode
1	solid phase of electrode
2	solution phase of electrode

7.1 Introduction

Increasing greenhouse gas (GHG) emissions from the fossil fuel-based energy systems and other factors have led to the development of renewable energy-based systems for electrical power production, as well as the development of smart cities that use stable smart grids with clean sources of energy. The availability of renewable energy sources like solar, wind and wave depends on the weather conditions and time of day, causing fluctuations, unavailability, and unpredictability of produced energy (Rosen and Farsi, 2022). Thus, the use of large-scale energy storage integrated with energy generation systems (water turbines, solar panels, wind turbines, etc.) has been considered to improve the reliability and overall usefulness of the energy systems.

Batteries are electrochemical storage systems that can be used in the electricity grid and various stationary and mobile applications to store electrical energy and then return it when it is in demand. Battery electrochemical-based storage systems have several advantageous attributes, including high efficiency, operation at various geographical locations scalability, low weight and mobility compared to many other electrical storage systems like pumped hydroelectric storage (Cho et al., 2015).

Increasing the use of renewable energy-based systems around the world leads to an increase in the use of battery storage systems to balance the intermittent and fluctuating electrical power production. For instance, in wind turbine and photovoltaic solar systems, the use of a battery system is necessary to store electrical energy when electricity production is higher than demand. The stored energy in the battery can be recovered later when there is insufficient wind speed and/or solar radiation to meet the electrical power demand. So, the use of battery technology results in load leveling and supporting the practical application of renewable energy-based electrical power production systems, as shown in Fig. 7.1.

Several battery technologies are well developed and commercial like lithium-ion batteries, lead acid batteries, and nickel metal hydride batteries, all of which are currently used for power system stabilization. A broad selection of batteries for electrical storage applications can be achieved through specification of the storage scale. For instance, lithium-ion battery technology is appropriate for small and medium scale storage applications like mobile phones and cars, whereas vanadium redox battery technology is more suitable for large-scale stationary applications. The lithium-ion battery is the most widely-used type of battery technology for mobile and portable applications due to its high efficiency and high energy density on a weight basis. However, the use of lithium-ion battery in stationary storage applications is limited due to its high cost. Further commercialization of battery technologies relies on lowering the cost per unit energy delivery, extending

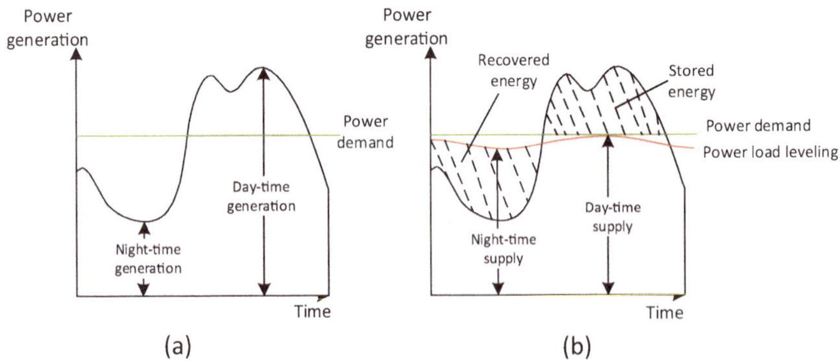

Fig. 7.1 Electrical power production profile from a renewable energy-based power plant. (A) Without battery storage, (B) with battery storage.

the lifetime and ensuring adequate supplies of the electroactive materials used in batteries.

Energy storage is becoming increasingly important. The annual increases in energy storage capacity by country are shown in Fig. 7.2 for the years 2015–20. It is seen that in 2020, 5 GW was added to the global battery storage capacity, and that by the end of 2020, the total installed global battery power storage capacity was about 17 GW (IEA, 2021). The total battery storage capacity grew by about 64% in 2020 compared to 2017. China and United States are leading in gigawatt-scale addition of battery storage capacity. Note that about two-third of the total added capacity in Fig. 7.2 is associated with utility-scale installations (IEA, 2021).

7.2 Integrated battery-based systems in transportation

There is increasing interest in the use of battery technology in transportation applications, such as aviation, road and marine vehicles and trains due to their low noise, high efficiency and low emissions. In particular, the use of lithium-ion battery cells in electric and hybrid electric cars is promising as a means to reduce significantly GHG emissions from the transportation sector.

Fig. 7.3 shows the existing and predicted carbon dioxide emissions from the global transportation for 2000–30 (IEA, 2022). The greatest emissions are seen to be associated with road passenger followed by road freight vehicles, followed by shipping and aviation. Policies and sustainable development plans

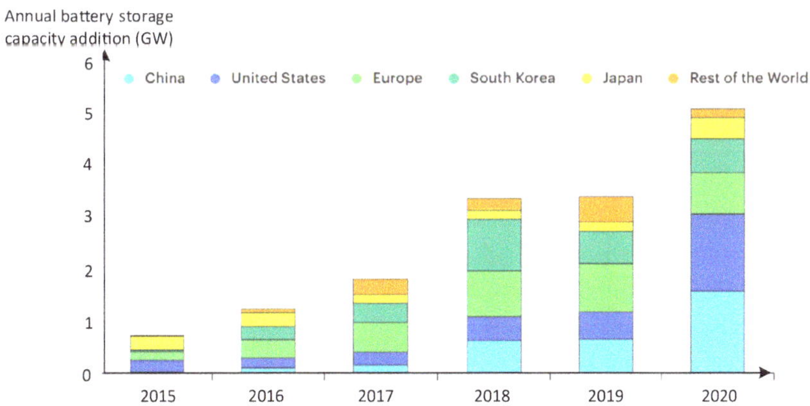

Fig. 7.2 Annual energy storage additions by country from 2015 to 2020. *(Modified from IEA, 2021. Energy Storage, IEA, Paris. Available from: https://www.iea.org/reports/energy-storage (Accessed 25 May 2022).)*

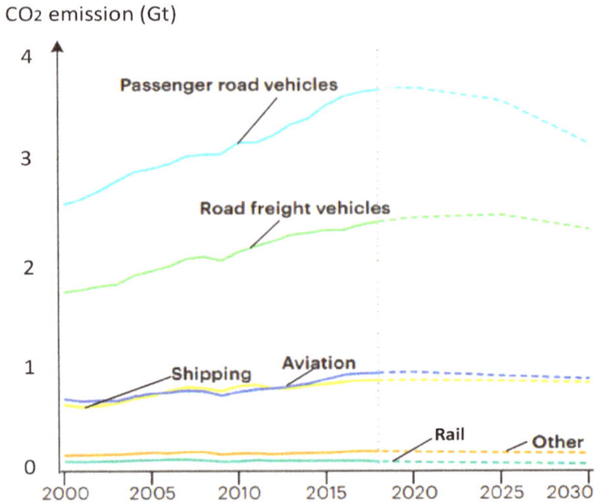

Fig. 7.3 Existing and predicted transportation sector annual carbon dioxide emissions for 2000–30. Other includes pipeline and nonspecified transport. *(Modified from IEA, 2022. Transport sector CO_2 emissions by mode in the Sustainable Development Scenario, 2000–2030. IEA, Paris. Available from: https://www.iea.org/data-and-statistics/charts/transport-sector-co2-emissions-by-mode-in-the-sustainable-development-scenario-2000–2030 (Accessed 25 May 2022).)*

are required to promote the use of low-carbon fuels to decarbonize road passenger and heavy-duty road freight vehicles as well as aviation and shipping, especially with the expected growth in transport demand. It is predicted that, by establishing such policies from 2018 to 2030 (represented by the dashed curves to the right of the vertical dotted line in Fig. 7.3), transport sector carbon dioxide emissions can be reduced by 20% to 5.7 Gt per year (IEA, 2022).

Reducing carbon dioxide emissions from marine transportation though diesel-based propulsion systems of ships requires strict regulations (Hoang, 2018). Numerous projects have aimed at replacing diesel-based propulsion with fuel cell-battery combined systems. For instance, in 2017–21, in the ReverCell project, proton exchange membrane (PEM) fuel cells with a total power output of 250 kW were combined with battery systems to drive hybrid ferries (Zero emissions, 2020). In addition, in 2021 a new project was initiated in which solid oxide fuel cells (SOFCs) with a total power output of 60 kW are integrated with a battery and internal combustion engine system to drive cruise ships (Horizon, 2020). Although fuel cell-battery-based power systems have been successfully implemented in marine projects, several challenges like high manufacturing costs, fuel storage challenges, maintenance and safety issues need to be addressed to stimulate wider use of fuel cell-battery technology in ships.

The development of electric aircraft has recently received increasing global interest to reduce cost per passenger and GHG emissions (Collins and McLarty, 2020). Several companies like Zunum Aero, Airbus, and Eviation are investigating battery-based short range passenger aircraft. The hybridization of alternating propulsion systems and associated flight performance is addressed by Ji et al. (2020). Also, the National Aeronautics and Space Administration (NASA) has invested notably in electric and hybrid electric aircraft research, and expects larger prototypes to be test-ready by 2025 (Jansen et al., 2019). Collins and McLarty (2020) proposed a fuel cell-gas turbine hybrid system for an aircraft that utilizes liquid hydrogen as fuel. They describe the geometry and material requirements for a fuel cell-gas turbine-battery hybrid system. The performance of the proposed system is analyzed for four commercial aircraft flight profiles, and optimized for maximum payload capacity (i.e., the maximum weight it can safely carry). The authors show that the high energy density of liquid hydrogen and the high efficiency of SOFCs decrease the fuel requirement by 70%, which offsets the weight of the onboard electric power unit. Furthermore, Seyam et al. (2021) propose a hybrid propulsion system for an aircraft in which a turbofan is integrated with a SOFC. Thermodynamic and parametric studies are carried out and the performance of the system is assessed through energy and exergy methods. Also, the use of various types of fuels such as methane, hydrogen, methanol and ethanol in numerous combinations is investigated as a means to replace kerosene, the conventional fossil-based fuel used in aircraft. The authors show that the integration of a SOFC with a conventional turbofan can improve system performance. Also, the use of a fuel mixture of 40% hydrogen and 60% ethanol increases the performance by 5% and decreases carbon emissions by 73%.

Much research has been reported on electric and hybrid aviation vehicles. For instance, actual European flight information has been collected by Kousoulidou and Lonza (2016) to compare the consumption of bio-kerosene and conventional kerosene in terms of carbon dioxide emissions. It was found that consuming 170 million tons of conventional kerosene results in emission of 400 million tons of GHG emissions by 2030. So, one European Union plan calls for the use of biofuels like hydrotreated vegetable oils (HVOs) and biomass-to-liquid (BTL) biojet fuel to reduce carbon dioxide emissions from the aviation transportation (Kousoulidou and Lonza, 2016).

Electric and hybrid electric vehicles can be classified into four main types: battery electric, fuel cell electric (or hydrogen electric), plug-in hybrid electric and hybrid electric (see Fig. 7.4). For battery electric vehicles, battery cells are the sole energy supply of the vehicle. Battery electric vehicles have large rechargeable batteries which can provide reasonable driving ranges, e.g., over 322 km (200 miles) of range via a 60–100 kWh battery pack (Oreizi, 2020). The battery can be charged from 10% to 80% in about 30 min with fast charging (Oreizi, 2020). Battery electric vehicles typically have higher efficiencies than hybrid electric vehicles. The plug-in hybrid electric vehicle is a combination of a hybrid vehicle and a battery electric vehicle. Such a vehicle uses both rechargeable batteries and a gasoline fuel tank to drive the vehicle. The battery can be recharged by plugging in or through regenerative braking. Also, the plug-in hybrid electric vehicle can operate solely on an electric power source that supplies the batteries. The hybrid electric vehicle is typically comprised of batteries and a gasoline fuel tank. Typically, the battery in hybrid cars is charged by on-board operations and cannot plug into the grid to recharge. Hybrid electric vehicles have much smaller battery capacity compared to plug-in hybrid electric vehicles. Fuel cell vehicles (hydrogen electric vehicles)

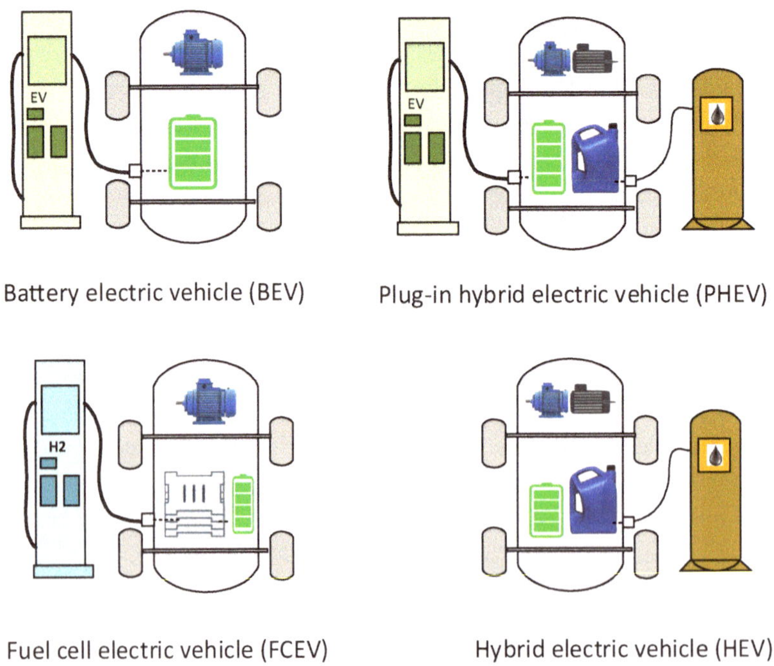

Fig. 7.4 Main types of electric and hybrid electric vehicles.

use hydrogen as a fuel for driving the vehicle. This type of electric vehicle is less common than the others. The PEM fuel cell is often used in hydrogen electric cars that use hydrogen. A battery pack can be used in such cars to supplement the fuel cell during high driving load conditions.

7.3 Case studies

Fuel cell and battery technologies are promising, often due to their high efficiencies and low or zero carbon dioxide emissions. However, these technologies are limited by low capacity and relatively slow dynamic response, especially for large-scale mobile applications like aircraft and ships. These limitations can be mitigated by the hybridization of fuel cells and batteries with conventional power supply systems. PEM fuel cell systems require a large fuel storage system because of the low volumetric energy density of hydrogen. This is recognized as one of the main drawbacks for aviation and marine propulsion applications. The SOFC, with its high fuel flexibility, can be a more attractive electrical generation system. In addition, in aviation and marine applications of fuel cell-battery systems, it is often recommended to use a fuel reformer which enables the use of natural gas or methanol fuels in the SOFC instead of using pure hydrogen. The reformer is a device that produces hydrogen from a hydrogen-containing source such as short-chain hydrocarbon fuels like methane, natural gas and ethanol.

Fuel cell and battery systems can drive electric and hybrid electric vehicles. In a fuel cell electric vehicle, a PEM fuel cell is often the main power supply and a lithium-ion battery set is often the auxiliary power unit. However, the car manufacturer Nissan has recently introduced an electric vehicle in which a SOFC-reformer and battery are used to drive the car. Ethanol is used as the input fuel to the car and is converted to hydrogen through reforming. Fig. 7.5 shows the fuel cell car developed by the Nissan Company (E-Bio fuel-cell, 2017). The electrical power from the fuel cell and battery is converted to mechanical energy by the motor and transmitted to the drive system. The batteries in fuel cell cars are charged either through regenerative braking (in which part of the generated energy is used to charge the batteries) or via the fuel cell during low load driving.

In all mobile applications of battery systems, including marine, aviation and road vehicles, thermal management of battery cells is an important factor for consideration in vehicle design. The battery thermal management system maintains the battery temperature within the desired operating range (typically 15–45°C). There are three main types of battery thermal management

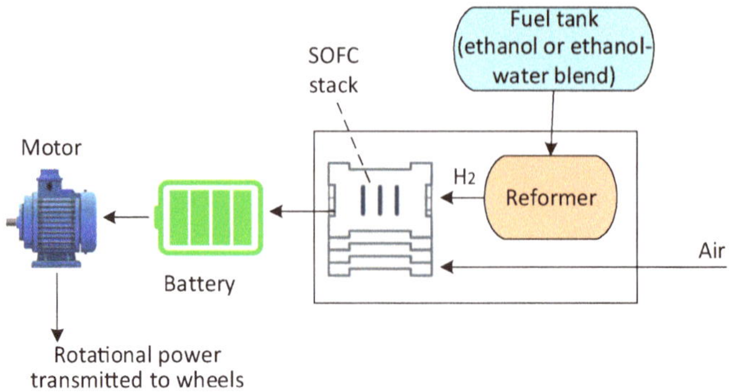

Fig. 7.5 Layout and energy conversion principle of the Nissan solid oxide fuel cell vehicle.

systems: air-based (Zhang et al., 2021; Lin et al., 2021), liquid-based (Kalaf et al., 2021; Rao et al., 2017), and phase change material-based. The latter category of battery thermal management system includes systems based on solid-liquid phase change (Chen et al., 2021; Malik et al., 2016), and liquid-vapor phase change (Al-Zareer et al., 2019; Wang et al., 2016). Chapter 5 describes comprehensively various types of battery thermal management systems.

In this chapter, the integration of some battery-based technologies is presented and the performance is discussed to see how it is affected by varying the operating conditions. In the first case study, an integration of a battery thermal management system with a fuel cell/battery electric vehicle is examined. In the second case study, an electric aircraft propulsion system including a SOFC, a lithium-ion battery and a gas turbine is described. Comprehensive modeling, analyses and assessments of each system are presented, providing an understanding of and insights into the development of new battery-based systems. The overall performances of the developed systems are discussed, and recommendations for further improvements are provided.

7.4 Case study 1: PEM fuel cell-assisted lithium-ion battery electric vehicle integrated with an air-based thermal management system

Global interest in electric vehicles has been growing, due in part to their zero-emission characteristics, high efficiency and low noise. Electric vehicles can be driven by batteries only, or another power source can be integrated with batteries like a fuel cell. In a battery-only electric vehicle, the electrical power recovered from the battery is directly used in the electrical motor, has allowing for high efficiency. However, such types of electric vehicles have disadvantages such as long charge duration and limited battery capacity (Wang and Fang, 2017). Thus, research and development efforts into integrated power sources for electrical vehicles like fuel cell–battery systems have been growing.

Another important issue in electric vehicles is thermal management of the battery packs. The performance and lifetime of battery cells, specifically lithium-ion batteries, are highly affected by operating temperature. Various thermal management systems—including air-based (Zhang et al., 2021), liquid-based (Kalaf et al., 2021), and phase change material-based (Chen et al., 2021; Al-Zareer et al., 2019)—are undergoing development and some are used commercially in electric vehicles. In this context, this case study describes an electric vehicle in which a proton exchange membrane (PEM) fuel cell is integrated with lithium-ion batteries for supplying electrical power for a vehicle. The battery cells are cooled by an air-based thermal management system.

Fig. 7.6 shows a fuel cell-battery electric vehicle and its main components. They include the following: PEM fuel cell stack, lithium-ion battery pack, high-pressure hydrogen tank, motor, power control unit, pressure reducer valve and fuel cell boost convertor. The PEM fuel cell stack is comprised of 360 cells arranged in series. These form the main power source of the electric vehicle. The hydrogen is stored in a high pressure tank (up to 70 MPa), with a sufficient hydrogen storage capacity to allow the electric vehicle to run between 500 and 600 km before refueling. Since, low-pressure hydrogen is required for proper function of the PEM fuel cell, one or more pressure reducing valves is used to depressurize the hydrogen before it enters the fuel cell stack.

A compact and high-efficiency boost converter is used to increase the fuel cell stack voltage so as to obtain an output with a higher voltage than the input. The motor is driven by electrical power generated by the PEM fuel cell stack and supplied by the lithium-ion battery pack, which contains 84 cells. The lithium-ion battery pack stores energy from the fuel cell during low load conditions as well as energy recovered from braking. The power control unit controls optimally both the fuel cell stack output under its operational conditions and the discharging and charging of the

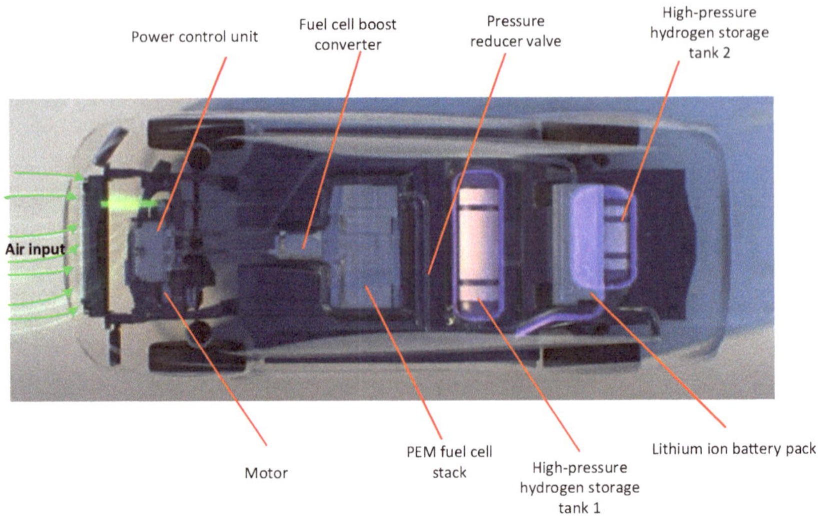

Fig. 7.6 Main components of a fuel cell/battery-assisted electric vehicle. *(Figure modified from Toyota, 2020. Toyota Europe Newsroom. Available from: https://newsroom.toyota.eu/introducing-the-all-new-toyota-mirai/ (Accessed 25 May 2022).)*

battery. There are different modes of operation depending on the driving load (see Fig. 7.7):

- During very low load driving, the battery alone can supply adequate power for the electric vehicle.
- During low load driving, the fuel cell drives the vehicle and excess power produced by the fuel cell is used to charge the battery.
- During high load driving, the fuel cell drives the vehicle and the battery assists the fuel cell stack in supplying power.

An increasing battery temperature due to the heat generation rate within the battery is an important issue that needs to be controlled. This is typically done through a battery thermal management system. In this case study, it is necessary to remove the generated heat from the lithium-ion battery as it adversely affects battery performance and can lead to battery failure. Battery thermal management systems are necessary to maintain the temperatures of batteries within a desired operating range (ideally between 15°C and 45°C) and to allow for safe operation. In this case study, an integration of the battery thermal management system with the fuel cell/battery electric vehicle is assessed in which the air that is drawn from the atmosphere for use in the PEM fuel cell is first circulated through the battery cells inside the pack. A part of the preheated air is then used in the fuel cell for electrical

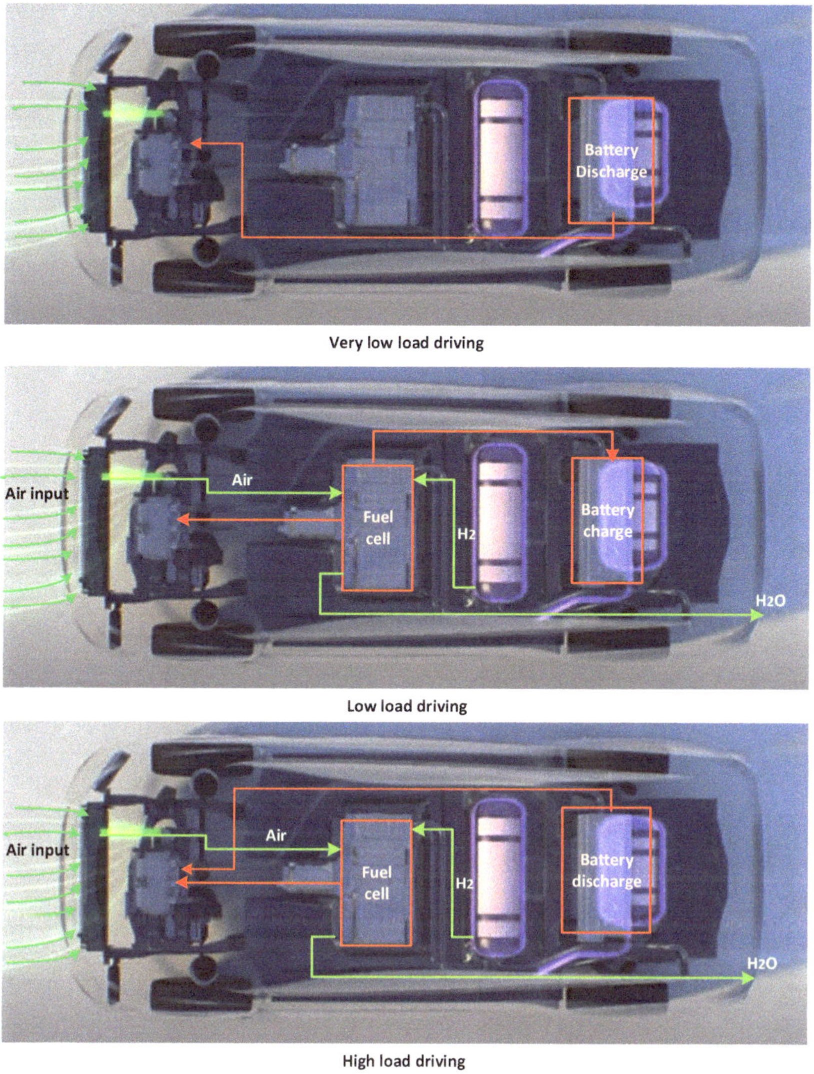

Fig. 7.7 Operation of fuel cell/battery electric vehicle for main driving load conditions.

power production (see Fig. 7.8). Note that preheating the air before it enters the fuel cell rather than directly using it in the fuel cell can potentially improve the performance of the fuel cell stack in terms of electrical power production and efficiency. The excess air and produced water exiting the PEM fuel cell stack are released to the environment. The other components of the electric vehicle (e.g., PEM fuel cell) are cooled via the radiator.

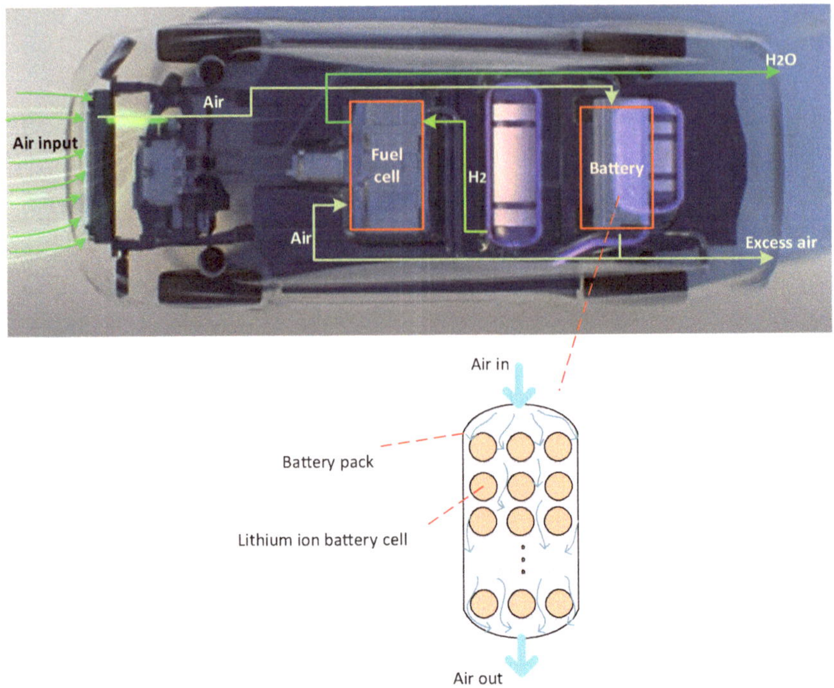

Fig. 7.8 Air-based thermal management of battery pack in the fuel cell/battery electric vehicle.

Fig. 7.8 shows an air-based thermal management of battery pack in the fuel cell/battery electric vehicle.

7.4.1 Electrochemical model of lithium-ion battery

Each battery cell in the pack includes negative and positive porous electrodes which are separated by a polymer separator. The porous electrodes consist of active an insertion electrode material (solid phase), a polymer electrolyte (solution phase) and a conductive filler that provides additional electronic conductivity. Carbon material and lithium manganese oxide ($LiMn_2O_4$) are used as active materials for negative and positive electrodes, respectively. A pseudo-homogeneous mixture is used for modeling the porous electrodes. Details on the thermophysical behavior and properties of porous electrodes and the polymer electrolyte during discharge and charge of the battery cell are provided elsewhere (Doyle et al., 1993, 1996; Fuller et al., 1994).

The battery isothermal model is developed for this case study in one dimension through the electrode/electrolyte's thickness. The thicknesses

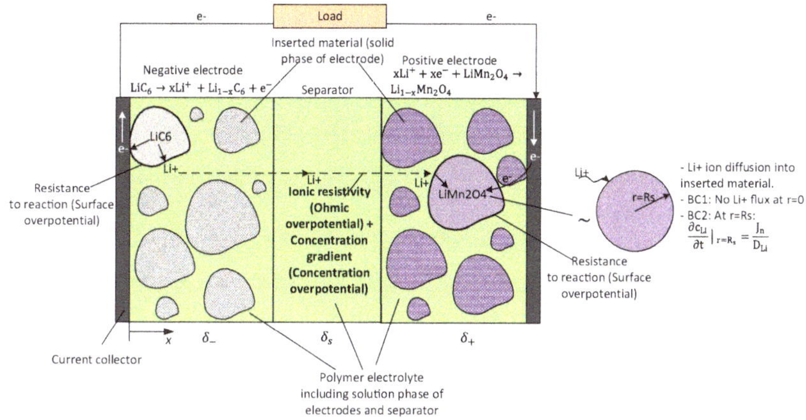

Fig. 7.9 Discharge operation of a lithium-ion battery consisting of composite porous negative and positive electrodes and a polymer separator. *From Farsi, A., Rosen, M.A., 2022b. PEM fuel cell-assisted lithium ion battery electric vehicle integrated with an air-based thermal management system. Int. J. Hydrog. Energy 47(84), 35810-35824.*

of the negative and positive electrodes are 100 and 180 μm, respectively, while the electrolyte thickness is considered to be 50 μm. Fig. 7.9 shows a schematic of the one-dimensional model for the lithium-ion battery cell. The model includes:

(i) ionic charge transport in the electrodes and electrolyte/separator to account for ohmic overpotential due to ionic transport,

(ii) material transport in the electrolyte to account for the effect of concentration on potential loss and ionic conductivity,

(iii) electronic current conduction in the electrode to account for the ohmic overpotential due to the passage of electronic current,

(iv) material transport within the spherical particles that form the electrodes, and

(v) Butler-Volmer electrode kinetics to account for the activation overpotential in the anodic and cathodic electrochemical reactions that occur at the electrodes.

7.4.1.1 Porous electrode

The variation with time of the salt concentration in the electrode solution phase (i.e., polymer electrolyte) of the porous electrode can be expressed based on a material balance for the lithium in the polymer/salt phase, as follows (Doyle et al., 1996):

$$\epsilon \frac{\partial c_s}{\partial t} = \left(\nabla \cdot (D_s \nabla c_s) + \bar{a} J_n (1 + t_+^0)\right) - \frac{i_2 \nabla t_+^0}{F}) \tag{7.1}$$

where c_s, ϵ, t, D_s, F, and $t_+{}^0$ respectively denote salt concentration, volume fraction of polymer, time, salt diffusion coefficient, Faraday constant, and transference number. The transference number indicates that the flow of ions creates an ionic current in an electrolyte. This means that each ion in the electrolyte contributes to the ionic current by carrying a fraction of the current in the solution. A more detailed description of the transference number is presented in Chapter 3 (Section 3.4). The terms i_2, $\bar{a}$, and J_n respectively denote ionic current density in the solution phase, specific interfacial area and the pore wall flux of lithium ions. These terms are related as follows:

$$\bar{a} = \frac{1}{F} \nabla \frac{i_2}{J_n} \tag{7.2}$$

where the ionic current density in the electrolyte phase of the electrode can be expressed as:

$$i_2 = -\kappa^{eff} \nabla \Phi_2 + \left(\frac{2\kappa RT}{F} \left(1 + \frac{\partial \ln \gamma_\pm}{\partial \ln c_s} \right) \right) \left(1 - t_+^0\right) \nabla \ln c_s \tag{7.3}$$

where κ^{eff} denotes the effective electrolyte ionic conductivity, R the universal gas constant, T the absolute temperature of the cell, $\gamma_\pm$ the activity coefficient and Φ_2 the electrical potential in the solution phase. Note in general that subscript 1 indicates the solid phase and subscript 2 the electrolyte phase of the composite electrode. The parameter κ^{eff} can be determined based on the ionic conductivity of electrolyte (κ_0) and the phase volume fraction (ϵ) as,

$$\kappa^{eff} = \kappa_0 \epsilon^\gamma \tag{7.4}$$

where γ is the Bruggeman coefficient, which is taken here to be 3.3 (Doyle et al., 1996). The ionic conductivity of the electrolyte depends on the cell temperature, salt concentration and solvent ratio of ethylene carbonate (EC) to dimethyl carbonate (DMC). The electrolyte used in this case study is comprised of 2 M lithium hexafluorophosphate ($LiPF_6$) salt in a 1:2 EC: DMC (by volume) solvent and a porous membrane comprised of polyvinylidene fluoride-hexafluoropropylene (PVDF-HFP). Fig. 7.10 shows the ionic conductivity of the electrolyte, based on Doyle et al. (1996). The conductivity values are measured experimentally and they change during discharge and charge as the electrolyte composition changes.

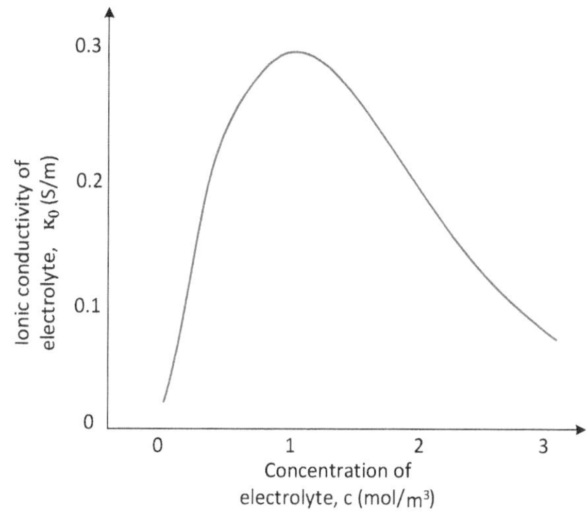

Fig. 7.10 Ionic conductivity change with the electrolyte concentration. *(Data from Doyle, M., Newman, J., Gozdz, A.S., Schmutz, C.N., Tarascon, J.M., 1996. Comparison of modeling predictions with experimental data from plastic lithium ion cells. J. Electrochem. Soc. 143(6), 1890–1903.)*

7.4.1.2 Insertion electrode material

The insertion electrode is made up of spherical particles with a radius of R_s. The lithium concentration in the electrode solid matrix can be defined based on the diffusion of lithium ions into the solid particles as follows:

$$\frac{\partial c_{Li}}{\partial t} = D_{Li}\left(\frac{\partial^2 c_{Li}}{\partial^2 r^2} + \frac{2}{r}\left(\frac{\partial c_{Li}}{\partial r}\right)\right) \tag{7.5}$$

where D_{Li} is the diffusion coefficient of lithium in the solid phase of the composite electrode and r is the dimension through the radius (radial direction) of the particles. One of the boundary conditions for Eq. (7.5) is no flux of lithium ions across at the center of solid particles (i.e., $\partial c_{Li}/\partial t=0$ at $r=0$). The other boundary condition, which relates the solid and polymer electrolyte phases of the porous electrode on the outer surface of electrode, is based on the relation between the lithium-ion flux into the particles and the rate of lithium-ion diffusion to the surface of the particles, i.e.,

$$\frac{\partial c_{Li}}{\partial t}\Big|_{r=R_s} = \frac{J_n}{D_{Li}} \tag{7.6}$$

7.4.1.3 Polymer separator

In the polymer separator, Eq. (7.1 is applied with $J_n = 0$ (no pore wall flux of lithium ions) and $\epsilon = 1$ (the entire separator is a polymer so the polymer fraction is 1). Thus, the variation with time of the salt concentration in the polymer separator can be expressed as:

$$\frac{\partial c_s}{\partial t} = (\nabla \cdot (D_s \nabla c_s)) - \frac{i_2 \nabla t_+^0}{F} \tag{7.7}$$

Ionic current density is conserved in the two regions of the separator and the electrolyte phase of porous electrode (i.e., i_2 (in separator) $= i_2$ (in electrolyte phase of electrode)). Based on Eq. (7.3), the electrical potential variation in the separator can be written as:

$$\nabla \Phi_2 = \frac{i_2}{\kappa^{eff}} + \frac{1}{\kappa^{eff}} \left(\frac{2\kappa RT}{F} \left(1 + \frac{2\kappa RT}{F} \right) \frac{\partial \ln \gamma_\pm}{\partial \ln c_s} \right) (1 - t_+^0) \nabla \ln c_s \tag{7.8}$$

The first term on the right side represents the ohmic overpotential due to the ionic resistivity of the electrolyte and the second term represents the concentration overpotential due to a concentration gradient within the polymer separator and solution phase of the porous electrode.

The boundary condition for the electrical potential at the polymer separator is no potential change in the solid phase, as follows:

$$\frac{\partial \Phi_1}{\partial x} \Big|_{x=\delta_- + \delta_s} = 0 \tag{7.9}$$

Other boundary conditions considered in the model of the lithium-ion battery cell are (i) a potential of zero for the current collector at the negative electrode and (ii) at the positive electrode current collector, the current density is specified as a finite constant through discharge and charge and zero through the open circuit duration.

7.4.1.4 Surface overpotential

The surface overpotential is the resistance to electrochemical reactions in the negative and positive porous electrodes. In fact, a driving force (activation energy) is required to induce the electrochemical reactions to occur. During discharge, the reactions at the negative and positive electrodes can be expressed as follows:

$$\text{Negative electrode: } LiC_6 \rightarrow xLi^+ + Li_{1-x}C_6 + e^- \tag{7.10}$$

$$\text{Positive electrode: } xLi^+ + xe^- + LiMn_2O_4 \rightarrow Li_{1-x}Mn_2O_4 \tag{7.11}$$

It can be assumed that the reactions at the electrodes follow Butler-Volmer kinetics. The general expression of the Butler-Volmer equation can be written as:

$$i = i_{0,1}\left(\exp\left(\frac{\alpha_{-}\eta_{s1}n_eF}{RT}\right) - \exp\left(\frac{-\alpha_{+}\eta_{s1}n_eF}{RT}\right)\right) \tag{7.12}$$

where i is the net current density. Note that, due to different rates for the positive and negative electrodes reactions, a net current density is produced in the cell. The terms α_{+} and α_{-} respectively are the transfer coefficients of reaction at the positive and negative electrodes. These are 0.5 for both reactions at the positive and negative electrodes. Also, $i_{0,1}$ is the exchange current density in the solid phase of electrode, which can be expressed as follows:

$$i_{0,1} = Fk_{+,1}^{\alpha_{+,1}}k_{-,1}^{\alpha_{-,1}}(c_{max} - c)^{\alpha_{+,1}}c^{\alpha_{+,1}} \tag{7.13}$$

Here, c_{max} and c are the solubility limit of the lithium salt and electrolyte concentration, respectively, and k_{+} and k_{-} are rate constants for the reactions occurring at the positive and negative electrodes, respectively.

The term η_{s1} in Eq. (7.12) is the local surface overpotential, which is equivalent to the difference between the positive/negative electrode potentials $(\Phi_1 - \Phi_2)$ and the open circuit potential $(E_{OCP,1})$. That is,

$$\eta_{s1} = E_{OCP,1} - (\Phi_1 - \Phi_2) \tag{7.14}$$

where Φ_1 and Φ_1 respectively are the electrical potentials of the solid and solution phases of the composite electrode, respectively. If a lithium foil is used for the negative electrode and it is considered the reference electrode, the open circuit potential for the lithium foil electrode is zero. For this case study, in which porous composite electrodes are used for both the negative and positive electrodes, the Butler-Volmer equation becomes:

$$j_n = k(c_{1,max} - c_{1,Li})^{0.5}c_{1,Li}^{0.5}\left[\exp\left(-\frac{F}{2RT}(\eta_s)\right) - \exp\left(\frac{F}{2RT}(\eta_s)\right)\right] \tag{7.15}$$

where $c_{1,max}$ and $c_{1,Li}$ are the maximum concentration of lithium in the solid phase of the composite electrode and the concentration of lithium in the solid phase of the composite electrode. The term k is the reaction rate constant for the charge transfer reaction at the electrode surface (i.e., the electrode-polymer separator interface). The reaction rate constant is k_{+} for the positive electrode surface and k_{-} for the negative electrode surface. The electrode surface overpotential η_s can be written as

$$\eta_s = E_{OCP} - (\Phi_1 - \Phi_2) \tag{7.16}$$

where $\Phi_{+/-}$ is the positive/negative electrode potential, which is the difference between the corresponding electrical potentials of all phases existing at the positive/negative electrode (i.e., $\Phi_{+/-} = (\Phi_1 - \Phi_2)$).

Doyle et al. (1996) proposed empirical correlations for the open circuit potential of the negative electrode composed of carbon material and for the positive electrode composed of lithium manganese oxide ($LiMn_2O_4$) for a very low discharge rate and a battery temperature of 20°C. That is,

$$E_{eq_-} = -0.16 + 1.32\exp(-3x) + 10\exp(-2000x) \tag{7.17}$$

$$\begin{aligned} E_{eq_+} &= 4.2 + 0.06\tanh[-14.56y + 8.61] - 0.03\left[\frac{1}{(1-y)^{0.5}} - 1.9\right] \\ &\quad - 0.16\exp\left(-0.05y^8\right) + 0.81\exp(-40(y-0.14)) \end{aligned} \tag{7.18}$$

where x and y are stoichiometric coefficients for the carbon electrode (LiC_6) and lithium manganese oxide ($LiMn_2O_4$).

The superficial current density in the electrode solid phase can be expressed by Ohm's law as follows:

$$i_1 = -\sigma\nabla\Phi_1 \tag{7.19}$$

where σ is the electronic conductivity, which indicates the resistance to electron transfer in the solid phase of the negative/positive electrodes, and $\nabla\Phi_1$ denotes the ohmic overpotential due to electron transfer.

From the conservation of current in two phases (i.e., $\nabla(i_1 + i_2) = 0$), the superficial current density is equal to the summation of the current densities at the solid and solution phases. That is,

$$i = i_1 + i_2 \tag{7.20}$$

Finally, the cell potential and current respectively can be written as

$$I_B = Ai \tag{7.21}$$

$$\begin{aligned} V_B &= \Phi_+ - \Phi_- \\ &= E_{OCP} + (\eta_{s+} + \eta_{s-}) + \left(\nabla\Phi_{2,+} + \nabla\Phi_{2,-} + \nabla\Phi_{2,sep}\right) \\ &\quad + (\nabla\Phi_{1,+} + \nabla\Phi_{1,-}) \end{aligned} \tag{7.22}$$

where A is the battery surface area, and Φ_+ and Φ_- respectively are the electrical potentials of the positive and negative electrodes. The term $(\eta_{s+} + \eta_{s-})$ denotes the surface overpotential at the positive and negative electrodes,

$(\nabla \Phi_{2,+} + \nabla \Phi_{2,-} + \nabla \Phi_{2,sep})$ is the concentration overpotential and ohmic overpotential due to ionic transfer in the electrolyte at the positive and negative electrodes and the separator, and $(\nabla \Phi_{1,+} + \nabla \Phi_{1,-})$ is the ohmic overpotential due to electron transfer in the positive and negative electrodes.

There are 84 lithium-ion battery cells in the pack which are arranged in 3 rows, each containing 28 cells. Therefore, the electrical power output from the battery pack during discharge is $(28V_B) \times (3I_B)$.

7.4.2 Heat generation flux in the lithium-ion battery

The heat generation flux in each cell of the battery pack, neglecting mixing and phase change effects, can be written as follows:

$$\dot{q}_{gen} = i_B(E_{OCP} - V_B) - T\Delta S\frac{I}{n_e F A_B} \tag{7.23}$$

where ΔS is the entropy change due to electrochemical reactions occurring in the cell and n_e is the number of electrons in the balanced electrochemical half reaction. The terms E_{OCP}, A_B, i and V_B respectively denote open circuit potential (equilibrium cell potential), battery cell area, cell current density and cell voltage, respectively. The heat generation rate within the battery is due to:

- Irreversible heat flux ($i_B(E_{OCP} - V_B)$) including heat generation due to ohmic, concentration and surface overpotentials (see Eq. 7.22).
- Reversible heat flux ($T\Delta S(1/n_e F A_B)$) due to entropy change during the electrochemical reactions in the electrodes.

Since the thermal conductivities of the components of the lithium-ion battery cell are relatively high, it can be assumed that the battery has a uniform temperature profile. In addition, an isothermal model of a lithium-ion battery provides useful information about the operation of the battery at various temperatures. In this study, therefore, a uniform and constant average temperature is considered for the battery (i.e., the battery is considered isothermal). So, the reversible heat term is zero, as there is no change in the entropy of the battery cell during isothermal operation.

7.4.3 Air-cooled battery thermal management unit

Based on the Toyota Mirai electric vehicle, there are 84 ($=N_B$) lithium-ion battery cells in the battery pack (Toyota, 2020). It is assumed that each of the cells, given the cells have a uniform and constant temperature, generate the same heat rate. The lithium-ion battery cells are cooled by the forced air that is drawn from the external environment by a fan and transferred to the battery pack. Thus, the heat generation rate in the lithium-ion battery can be expressed as follows:

$$A_B \dot{q}_{gen} = \dot{N}_{air}\left(\overline{h}_{air,out} - \overline{h}_{air,in}\right) + mc_p \frac{dT}{dt} \tag{7.24}$$

where $\overline{h}_{air,in}$ and $\overline{h}_{air,in}$ respectively are the molar enthalpies of air at the inlet and outlet of the battery pack, and $\dot{N}_{air}$ is molar flow rate of air. Also, m is the mass of the battery cell and c_p is the specific heat at constant pressure of the battery, which is almost constant during battery operation. In Eq. (7.24):

- The term $\dot{N}_{air}\left(\overline{h}_{air,out} - \overline{h}_{air,in}\right)$ is the portion of generated heat rate within the battery that is absorbed by the air cooling medium.
- The term $mc_p dT/dt$ denotes the portion of the generated heat that leads to an increase in the internal energy of the battery. Here, this term is zero, since a uniform and constant temperature (isothermal condition) is considered for the lithium-ion battery.

It is assumed that sufficient air is input to the electric vehicle for cooling of the battery cells so that the air output temperature is about 5°C lower than the battery temperature. Fig. 7.8 shows the battery thermal management system using forced air for cooling the lithium-ion battery cells. There is an adjustable inlet port for air which allows for higher input rates of air at higher battery temperatures. Part of the preheated air is directed to the cathode side of the PEM fuel cell stack where it reacts with hydrogen and produce electrical energy. A portion of the preheated air leaving the battery pack is transferred to the PEM fuel cell for reaction with hydrogen and the rest can be used to cool other components of the vehicle or released to the environment.

The input parameters for the lithium-ion battery model are listed in Table 7.1, along with their values. The parameter C is the rate of charge or discharge. For instance, where the battery capacity is 1 Ah, a charging current of 1C can completely charge a battery in 1 h, and a discharging current of 1C can completely discharge a fully charged battery in 1 h. The data listed in Table 7.1 are taken from relatively old studies (i.e., Doyle et al., 1993, 1996; Fuller et al., 1994; Farsi and Rosen, 2022b), however they are still valid and used in current studies.

7.4.4 PEM fuel cell stack

Fig. 7.11 shows a polymer electrolyte membrane (PEM), sometimes called a proton exchange membrane, fuel cell stack, and its working principle. It is

Table 7.1 Design input parameters for isothermal model of the lithium-ion battery cell.

Parameter	Value
1C discharge current density, i	$12\,A/m^2$
Solid phase lithium diffusivity for negative electrode, $D_{li(-)}$	$3.9\times10^{-14}\ m^2/s$
Solid phase lithium diffusivity for positive electrode, $D_{li(+)}$	$1\times10^{-13}\ m^2/s$
Particle radius negative electrode, $R_{s(-)}$	$1.25\times10^{-5}\ m$
Particle radius positive electrode, $R_{s(+)}$	$8\times10^{-6}\ m$
Battery temperature, T	25°C
Electrolyte phase volume fraction for positive electrode, ϵ_+	0.63
Electrode phase volume fraction for positive electrode, $1-\epsilon_+$	0.297
Initial electrolyte salt concentration, $c_{2,\,0}$	$2050\,mol/m^3$
Electrolyte phase volume fraction for negative electrode, ϵ_-	0.503
Solid phase volume fraction for negative electrode, $1-\epsilon_-$	0.471
Transport number, ${t_+}^0$	0.363
Diffusion coefficient for free stream electrolyte, D	$7.5\times10^{-11}\ m^2/s$
Maximum solid phase concentration for negative electrode, $c_{1,\,max}$	$26{,}390\,mol/m^3$
Initial concentration for negative active electrode material, $c_{1,\,0(-)}$	$14{,}880\,mol/m^3$
Initial concentration for positive active electrode material, $c_{1,\,0(+)}$	$3920\,mol/m^3$
Electrolyte salt concentration, c_s	$1000\,mol/m^3$
Solid phase conductivity for negative electrode, σ	$100\,S/m$
Reference exchange current density for negative electrode, ${i^{ref}}_{0(-)}$	$11.71\,A/m^2$
Reference exchange current density for positive electrode, ${i^{ref}}_{0(+)}$	$11.14\,A/m^2$
Reaction rate coefficient for positive electrode, α_+	0.5
Reaction rate coefficient for negative electrode, α_-	0.5
Bruggeman coefficient, γ	3.3
Discharge duration, t_{disch}	1800 s
Open circuit duration, t_{ocp}	200 s
Charge duration, t_{ch}	1800 s
Length of negative electrode, δ_-	$1\times10^{-4}\ m$
Length of separator, δ_s	$5\times10^{-5}\ m$
Length of positive electrode, δ_+	$1.80\times10^{-4}\ m$
C-rate for the parametric study, C	2

Data from Doyle, M., Fuller, T.F., Newman, J., 1993. Modeling of galvanostatic charge and discharge of the lithium/polymer/insertion cell. J. Electrochem. Soc. 140(6), 1526–1533. Doyle, M., Newman, J., Gozdz, A.S., Schmutz, C.N., Tarascon, J.M., 1996. Comparison of modeling predictions with experimental data from plastic lithium ion cells. J. Electrochem. Soc. 143(6), 1890–1903. Fuller, T.F., Doyle, M., Newman, J., 1994. Simulation and optimization of the dual lithium ion insertion cell. J. Electrochem. Soc. 141(1), 1–10.

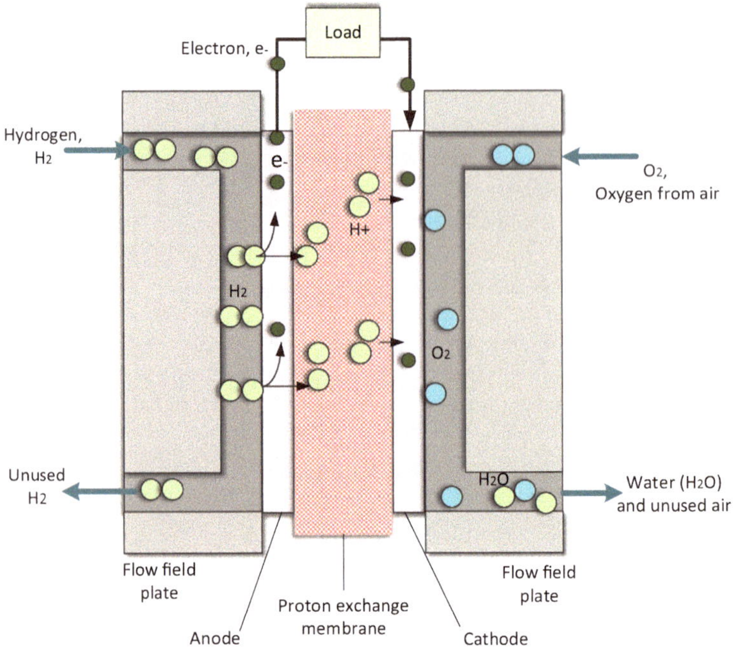

Fig. 7.11 Working principle of a proton exchange membrane (PEM) fuel cell. From Farsi, A., Rosen, M.A., 2022b. PEM fuel cell-assisted lithium ion battery electric vehicle integrated with an air-based thermal management system. Int. J. Hydrog. Energy 47(84), 35810–35824.

comprised of an anode, a cathode and a proton conductive membrane (an electrolyte). The hydrogen from the hydrogen storage tank is fed to the anode and oxidized at this electrode according to Eq. (7.25). Also, atmospheric air is drawn into the electric vehicle and introduced to the cathode side of the PEM fuel cell, where it is reduced according to Eq. (7.26). At the anode, hydrogen atoms dissociate to electrons and protons in the presence of a catalyst. The electrons move toward the cathode through the external circuit. The protons migrate through the proton exchange membrane toward the cathode where they react with oxygen and the electrons that were produced at anode to produce water and heat. The produced water and unused air leave the fuel cell from the cathode side and the unconsumed hydrogen from the anode side of the PEM fuel cell. The half-reactions at the anode and cathode and the overall cell reaction in the PEM fuel cell respectively can be written as follows:

$$\text{Anode: } H_2 \rightarrow 2H^+ + 2e^- \tag{7.25}$$

$$\text{Cathode: } 2H^+ + 2e^- + 0.5O_2 \rightarrow H_2O \tag{7.26}$$

$$\text{Overall: } H_2 + 0.5O_2 \rightarrow H_2O \tag{7.27}$$

In order to determine the electrical power production from the PEM fuel cell, the fuel cell voltage is determined based on the activation, concentration and ohmic overpotentials that exist during operation of the fuel cell.

The input flow rates of air and hydrogen can be varied depending on the speed of the electric vehicle. The output voltage from the PEM fuel cell can be written as:

$$V_{FC} = E_{OCP,FC} - \left(\left(\eta_{a,a} + \eta_{a,c} \right) + \left(\eta_{c,a} + \eta_{c,c} \right) + \eta_{ohm} \right) \tag{7.28}$$

where $E_{OCP,\ FC}$ is the open circuit potential of the PEM fuel cell, $\eta_{a,\ a}$ and $\eta_{a,\ c}$ are the activation overpotentials at anode and cathode respectively, $\eta_{c,\ a}$ and $\eta_{c,\ c}$ are the concentration overpotentials at the anode and cathode respectively and η_{ohm} is the ohmic overpotential.

7.4.4.1 Open circuit potential of PEM fuel cell

The open circuit potential of a fuel cell, $E_{OCP,\ FC}$, represents the electrical potential when the fuel cell operates at thermodynamic equilibrium. For the PEM fuel cell, $E_{OCP,\ FC}$ can be determined from the modified Nernst equation:

$$E_{OCP,FC} = E_{rev} + \frac{RT}{2F} \ln \left(P_{H_2} P_{O_2}{}^{0.5} \right) \tag{7.29}$$

where E_{rev} is the reversible cell potential, which denotes the Gibbs free energy that could be converted directly to electrical work without any losses. The terms R, F, and T are universal gas constant, the Faraday constant and the fuel cell absolute temperature, respectively. Also, P_{H_2} and P_{O_2} are the partial pressures of hydrogen at the anode/gas interface and oxygen at the cathode/gas interface.

The determination of the hydrogen and oxygen partial pressures requires mass transfer calculations in which average values for the direction of gas flow are typically used. Some parameters, like oxygen/hydrogen gas consumption rates, as well as chemical reactions in the fuel cell, change the partial pressures of reactants during operation. In this case study, the atmospheric air containing oxygen (molar fraction of oxygen in air is 0.21) is drawn in from the atmosphere, circulated through the battery cells and input to the PEM fuel cell. Since only a small fraction of the total supplied air is consumed at the cathode, the air velocity is approximately constant and consequently the oxygen partial pressure can be approximated using the log-mean average of the inlet and outlet oxygen partial pressures (Amphlett et al., 1995), as follows:

$$P_{O_2,ave} = \frac{P_{O_2,in} - P_{O_2,out}}{\ln\left(\frac{P_{O_2,in}}{P_{O_2,out}}\right)} \tag{7.30}$$

Here, the inlet oxygen pressure $P_{O_2,in}$ is equal to the pressure of air multiplied by 0.21. The cathode outlet product is a mixture of the produced water and the unconsumed air, and these are released to the atmosphere. So, the partial pressure of oxygen in this outlet mixture can be written as:

$$P_{O_2,out} = \frac{\dot{N}_{O_2,out}}{\dot{N}_{O_2,out} + \dot{N}_{H_2O,out}}(P_{atm}) \tag{7.31}$$

where

$$\dot{N}_{O_2,out} = \dot{N}_{O_2,in} - \dot{N}_{O_2,cons} \tag{7.32}$$

Also, the air utilization factor can be defined as follows:

$$U_{O_2} = \frac{\dot{N}_{O_2,cons}}{\dot{N}_{O_2,in}} \tag{7.33}$$

where the consumed oxygen mole flow rate $\dot{N}_{O_2,cons}$ can be found using the Faraday law and the stoichiometry coefficient of the overall reaction in the PEM fuel cell. Based on the Faraday law, the molar consumption rate of hydrogen at the anode can be determined as

$$\dot{N}_{H_2,consumed} = \frac{i.N_{FC}.A_{FC}}{2F} \tag{7.34}$$

where i is the net exchange current density of the fuel cell, N_{FC} the number of fuel cells in the stack, A_{FC} the active area of the cell, and F is the Faraday constant. From the reaction stoichiometry equation (Eq. 7.27), the rates of oxygen consumption and water production at the cathode are both half of the hydrogen consumption rate (i.e., $\dot{N}_{O_2,cons} = \dot{N}_{H_2O,out} = 0.5\dot{N}_{H_2,cons}$).

For estimation of hydrogen gas partial pressure, it can be assumed that the consumption of hydrogen contributes a relatively large fraction of total flow (the excess unconsumed hydrogen is small). Therefore, an average mean provides a good approximation of the hydrogen partial pressure, as follows (Amphlett et al., 1995):

$$P_{H_2,ave} = \frac{P_{H_2,in} + P_{H_2,out}}{2} \tag{7.35}$$

where the outlet pressure of hydrogen, $P_{H_2,out}$, from the anode can be written as

$$P_{H_2,out} = \frac{\dot{N}_{H_2,out}}{\dot{N}_{H_2,in}} P_{H_2,in} \tag{7.36}$$

The outlet molar flow rate of hydrogen is equal to the inlet molar flow rate of hydrogen minus the molar flow rate of hydrogen that is consumed by the fuel cell. That is,

$$\dot{N}_{H_2,out} = \dot{N}_{H_2,in} - \dot{N}_{H_2,cons} \tag{7.37}$$

Also, the fuel (hydrogen) utilization factor can be defined as:

$$U_{H_2} = \frac{\dot{N}_{H_2,consumed}}{\dot{N}_{H_2,in}} \tag{7.38}$$

Amphlett et al. (1995) proposed the following empirical equation for the reversible potential of a PEM fuel cell:

$$E_{rev} = E^{\circ}_{rev} + \left(T - T_{ref}\right)\frac{\Delta S^{\circ}}{2F} \tag{7.39}$$

where E°_{rev} is the reversible theoretical cell voltage at standard conditions, which is equal to 1.23 V. The term $(T - T_{ref})\Delta S^{\circ}/2F$ accounts for changes in the reversible voltage at temperatures that deviate from the standard reference temperature (T_{ref}) at standard conditions ($T = 25°C$ and $P = 1$ atm). Based on Amphlett et al. (1995), the term $\Delta S^{\circ}/2F$ can be approximated as -0.9×10^{-3} J mol/(K C).

7.4.4.2 Overpotentials in PEM fuel cell

The *activation overpotential* at the anode and cathode of a fuel cell denotes the rate of electrochemical reaction that is limited by the activation energy barrier to charge transfer between the electrode and electrolyte. The kinetics of the reactions occurring at the anode and cathode are assumed to follow the Butler-Volmer equation, which gives the current density resulting from those reactions. When the activation overpotential is large and negative at the cathode and is positive at the anode, the Butler-Volmer equation simplifies to the Tafel equation. For the PEM fuel cell, therefore, the cathodic and anodic activation overpotentials respectively can be expressed by the Tafel approximation, as follows:

$$\eta_{a,a} = \frac{RT}{\alpha_a F} \ln\left(\frac{i}{i_{0,a}}\right) \tag{7.40}$$

$$\eta_{a,c} = \frac{RT}{\alpha_c F} \ln\left(\frac{i}{i_{0,c}}\right) \tag{7.41}$$

where $i_{0,a}$ and $i_{0,c}$ are the exchange current densities at the anode and cathode, respectively, and α_a and α_c are the transfer coefficients of reaction at anode and cathode, respectively. The exchange current densities at the anode and cathode respectively can be determined as follows (Thampan et al., 2001):

$$i_{0,a} = \gamma_M \exp\left[-\frac{\Delta G_{C,a}}{R}\left(\frac{1}{T} - \frac{1}{T_{ref}}\right)\right] i_{0,a}^{ref} \tag{7.42}$$

$$i_{0,c} = \gamma_M \exp\left[-\frac{\Delta G_{C,c}}{R}\left(\frac{1}{T} - \frac{1}{T_{ref}}\right)\right] i_{0,c}^{ref} \tag{7.43}$$

where γ_M, ΔG_C and i_0^{ref} denote the electrode surface roughness factor, the activation free energy and the exchange current density at the reference state (T_{ref}, P_{ref}). Values for these parameters, as well as design parameters for PEM fuel cell, are listed in Table 7.2.

A *concentration overpotential* occurs in a PEM fuel cell when the surface coverage of oxygen bubbles at high current density prevents the hydrogen flow at high rate, consequently deterring the reaction. Experimental studies have found that concentration overpotential is typically much smaller than the other overpotentials, and it is often neglected. The concentration (or diffusion) overpotential accounts for concentration and transport losses. Mass flow through the anode and cathode is modeled using Fick's law, resulting in binary diffusion coefficients for the reactants. The limiting diffusion rate at relatively high current densities can be determined by combining the Nernst expression with Fick's law. This permits determination of the electrical potential drop caused by excess products, which block the catalyst sites for the reactants. Therefore, the concentration overpotential can be determined based on the current density (i) and limiting current density (i_L) at the anode and cathode respectively, as follows:

$$\eta_{c,a} = \frac{RT}{F} \ln\left(1 - \frac{i_a}{i_L}\right) \tag{7.44}$$

Table 7.2 Design input parameters for PEM fuel cell stack model.

Parameter	Value
Membrane thickness, δ_m	0.0254 cm
Cathode thickness, δ_c	0.008 cm
Anode thickness, δ_a	0.008 cm
Activation free energy for anode, $\Delta G_{C,a}$	29 kJ/mol
Activation free energy for cathode, $\Delta G_{C,c}$	66 kJ/mol
Current density, i	0.7 A/cm^2
Atmospheric pressure, P_{atm}	101 kPa
Air utilization factor, U_{O_2}	0.2
Hydrogen utilization factor, U_{H_2}	0.8
Hydrogen gas inlet pressure, $P_{H_2,\ in}$	135 kPa
Air inlet pressure, $P_{air,\ in}$	101 kPa
transfer coefficients of reaction at anode, α_a	0.7
Transfer coefficients of reaction at cathode, α_c	1.7
electrode surface roughness factor, γ_M	47
Exchange current density at the reference state for anode, $i_{0,\ a}^{ref}$	1×10^{-4} A/cm^2
Exchange current density at the reference state for cathode, $i_{0,\ c}^{ref}$	1×10^{-9} A/cm^2
Electrical resistivity of anode, ρ_a	16×10^{-3} Ωcm
Electrical resistivity of cathode, ρ_c	43.1×10^{-6} Ωcm
Effective water content of the membrane, λ	12.5
Active surface area of fuel cell, A	100 cm^2
Number of fuel cells in stack, N_{FC}	360
Inverter efficiency, η_{inv}	95%
Low heating value of hydrogen, $\overline{LHV}_{H_2}$	240,000 kJ/kmol
Reference pressure, P_{ref}	101 kPa
Reference temperature, T_{ref}	25°C

$$\eta_{c,c} = \frac{RT}{F}\ln\left(1-\frac{i_c}{i_L}\right) \tag{7.45}$$

For the PEM fuel cell, the limiting (maximum) current density is often in the range of 500–1500 mA/cm^2 (Corrêa et al., 2004; Gurau et al., 2000). The concentration overpotential term is included in the model described here. The maximum value of 1500 A/cm^2 is assumed for i_L since its effect can be significant at very high current densities. However, for a moderate operating range of current density, which is considered in this case study, the contribution of the concentration overpotential term is almost nil.

The *ohmic overpotential* includes electronic and ionic contributions from the cathode, anode and membrane of the fuel cell. The electronic resistance is the resistance of the electrically conductive cell components to the flow of electrons, while the ionic resistance is due to the flow of protons through the membrane. The ohmic overpotential in a PEM fuel cell can be written as:

$$\eta_{ohm} = (R_a + R_c + R_m)I \tag{7.46}$$

where R_a, R_a, and R_a are the resistances of the anode, cathode and membrane, respectively. The anodic and cathodic resistances, and the membrane resistivity, respectively can be determined as follows:

$$R_a = \frac{\rho_a \delta_a}{A} \tag{7.47}$$

$$R_c = \frac{\rho_c \delta_c}{A} \tag{7.48}$$

$$R_m = \frac{\delta_m}{\sigma_m A} \tag{7.49}$$

Here, ρ, δ, A, and σ_m denote the electrical resistivity of each electrode, length of each electrode, active area of the cell and proton conductivity of the membrane, respectively. The conductivity of the membrane, σ_m, for a Nafion membrane can be estimated as (Mann et al., 2000):

$$\sigma_m = \frac{\left[\lambda - 0.634 - 3\left(\frac{i}{A}\right)\right] \exp\left(4.18\left(\frac{T-303}{T}\right)\right)}{181.6 + \left[1 + 0.03\left(\frac{i}{A}\right) + 0.062\left(\frac{T}{303}\right)^2\left(\frac{i}{A}\right)^{2.5}\right]} \tag{7.50}$$

where λ is a semiempirical parameter representing the effective water content of the membrane and T is the PEM fuel cell absolute temperature. The PEM fuel cell model input parameters are presented in Table 7.2.

The output electrical power density and electrical power output of the PEM fuel cell stack, respectively, can be written as:

$$Pd = V_{FC} i \tag{7.51}$$

$$\dot{W}_{FC} = N_{FC} A(Pd) \tag{7.52}$$

where N_{FC} is the number of fuel cells in the stack. The PEM fuel cell energy efficiency can be expressed as the ratio of the electrical power output of the PEM fuel cell stack ($\dot{W}_{FC}$) multiplied by inverter efficiency (η_{inv}), which represents the useful output, to the molar low heating value of the input hydrogen ($\overline{LHV}_{H_2}$), which represents the fuel cell input. That is,

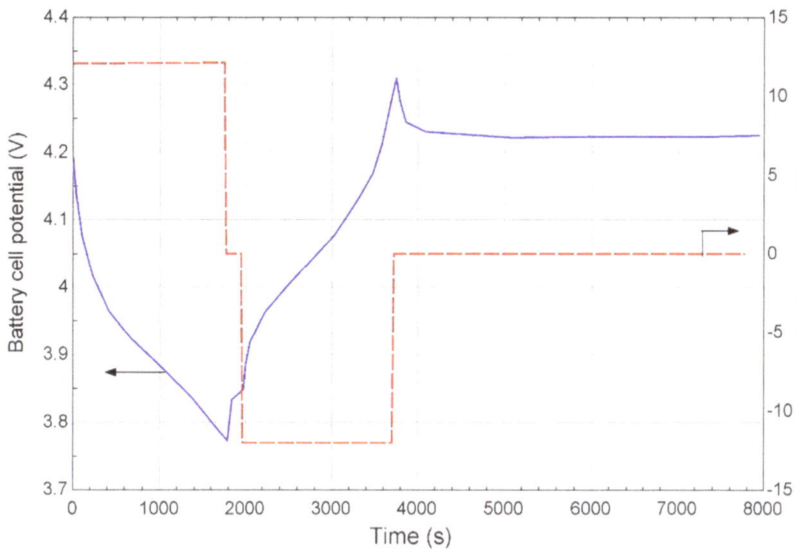

Fig. 7.12 Voltage and current of lithium-ion battery cell during discharge (0–1800 s) open circuit (1800–2000 s) and charge (2000–3800 s).

$$\text{Energy efficiency of PEM fuel cell} = \frac{\eta_{inv}\dot{W}_{FC}}{\dot{N}_{H_2,in}\overline{LHV}_{H_2}} \tag{7.53}$$

7.4.5 Results and discussion

In this section the results of the electrochemical assessment in the case study for the lithium-ion battery and the PEM fuel cell are presented. The isothermal model of lithium-ion battery is utilized via Comsol multiphysics software. The average results of the heat generation rate in the lithium-ion battery are transferred to engineering equation solver (EES) where the PEM fuel cell stack is modeled. The isothermal model of the battery helps in understanding the battery cell operation at various temperatures. In addition, it assists in designing the battery thermal management system and in specifying the temperature range at which the battery should operate. If battery operation is strongly affected by temperature, a precise temperature control unit is required for the battery thermal management system. In the isothermal model of the battery, all of the heat generated in the battery is lost through convection to the cooling medium (in this case forced air). The simulation results presented in this section for the lithium-ion battery

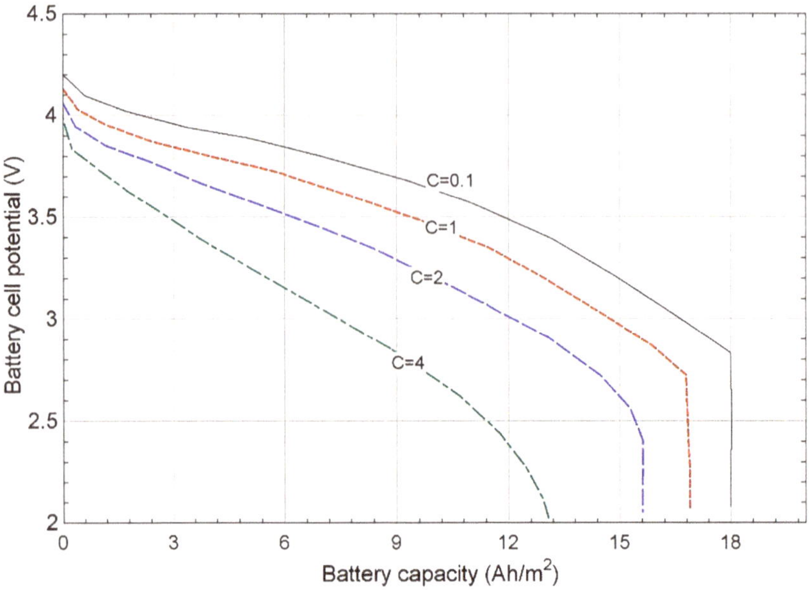

Fig. 7.13 Variation of battery cell potential with discharge capacity of the battery cell for four discharge rates (C-rates).

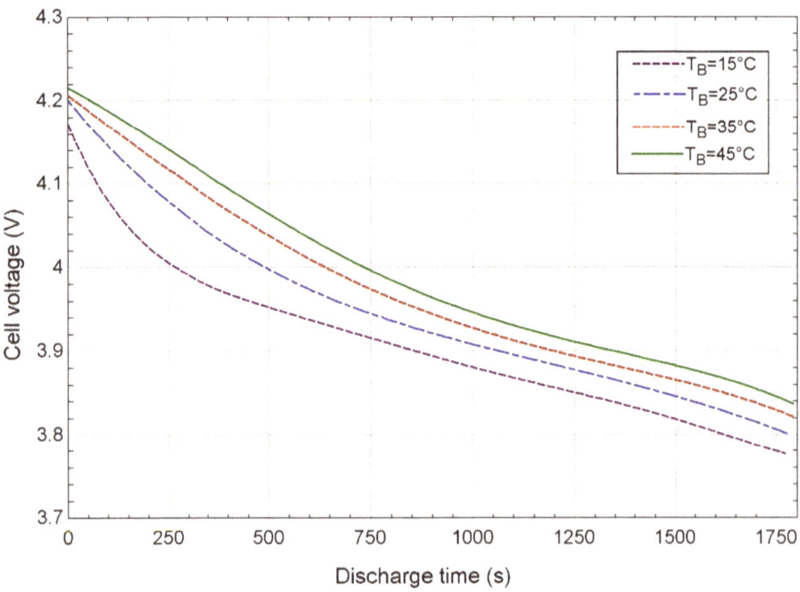

Fig. 7.14 Effect of lithium-ion battery temperature on the cell potential during discharge.

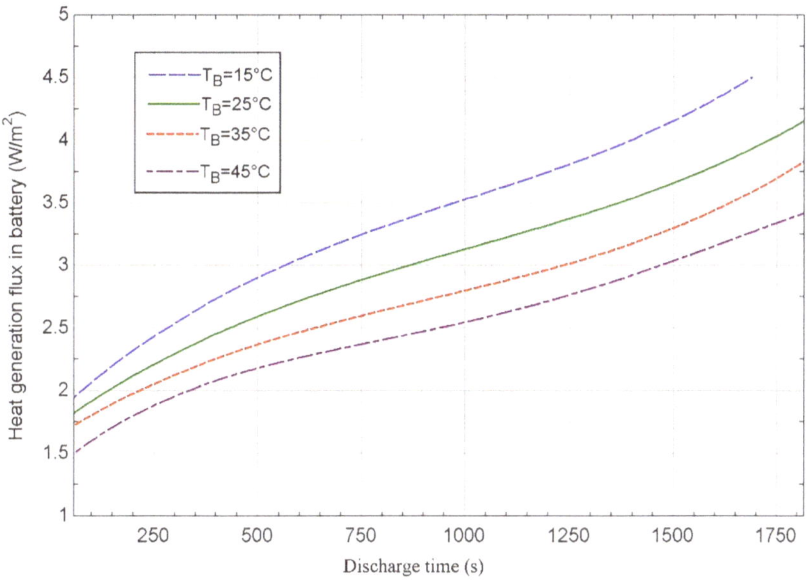

Fig. 7.15 Variation of heat generation flux in the lithium-ion battery with discharge time at various battery cell temperatures, for a discharge rate of 2C.

are for a discharge rate of $12\,A/m^2$. In the remainder of this section, the results obtained from applying the electrochemical models for the lithium-ion battery and PEM fuel cell are described.

Fig. 7.12 shows a single discharge-charge cycle for the lithium-ion battery cell. The charge and discharge durations are 1800 s each and the open circuit duration is 200 s. During the charge and discharge operation of the battery, the cell potential experiences a loss due to ohmic, surface and concentration overpotentials. During the open circuit (no current) condition, when the battery cell is at rest, the battery potential is observed to increase. This potential increase in the open circuit duration is associated with the portion of cell potential that is lost due to ohmic and concentration overpotentials during charge/discharge operations. A positive sign for the current during discharge indicates that current is generated by the battery cell during discharge, while a negative sign indicates current is applied to the battery cell during charge.

Fig. 7.13 shows the variation of the lithium-ion battery cell potential during discharge with the cell capacity for several rates of discharge (C-rates). It is seen that, with increasing C-rate, the maximum attainable capacity decreases. This means that, when the battery is discharged at a higher rate (e.g., fast discharging), the cell produces a lower current density over the

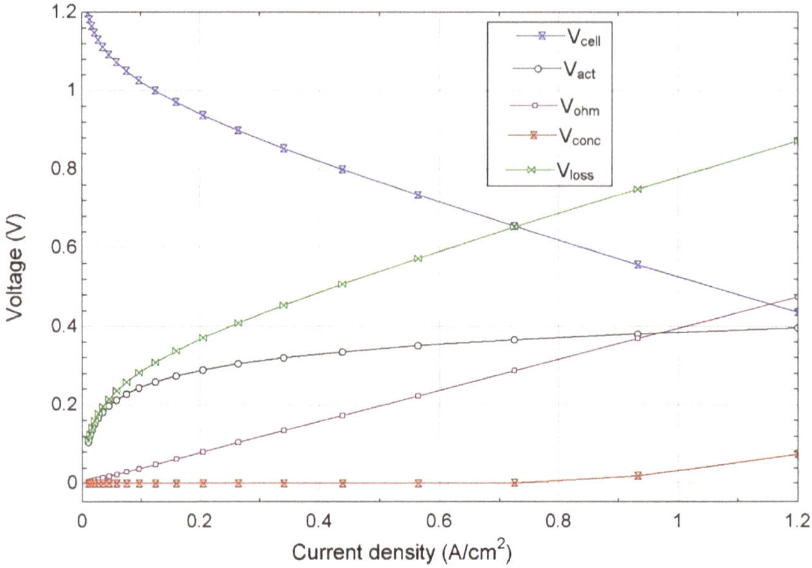

Fig. 7.16 Variations of cell potential and potential losses in the cell including ohmic, concentration and activation overpotentials with the applied current density.

duration of cell operation. In other words, the effective capacity of battery cell is reduced when discharge occurs at relatively high rates. For instance, at a discharge rate of 4C, the battery consumes about 40% of its capacity before reaching 3.2V.

Also, Fig. 7.14 shows the effect of battery temperature on the cell potential during discharge. It is seen cell potential increases as battery temperature increases. This is because of the better ionic conductivity and salt diffusion at relatively higher battery operating temperatures.

Fig. 7.15 shows the variation in heat generation flux in the lithium-ion battery cell with discharge time for several battery operating temperatures. As stated before, in this case study, in which isothermal operation of the battery is considered, the heat generation within the battery is comprised of only irreversible heat (the reversible heat rate component is zero). So, according to Eq. (7.23), Fig. 7.15 presents the difference between the cell open circuit potential and real potential multiplied by the current density. It is seen from this figure that the heat generation rate in the battery increases when the battery operates at relatively lower temperatures. This is mainly due to the lower ionic conductivity and larger ohmic loss at relatively lower temperatures, which leads to a higher heat generation rate. From the

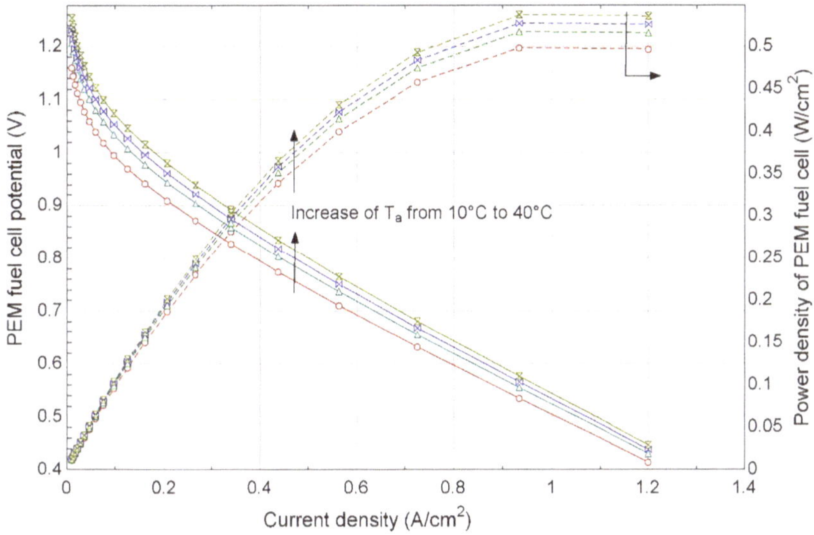

Fig. 7.17 Variation of PEM fuel cell potential and produced power density with current density of the fuel cell at various temperatures of the air input to the fuel cell.

assumption of uniform and constant temperature of the lithium-ion battery cells, it can be concluded that all of the heat generated within the battery is transferred to the cooling air before it enters the PEM fuel cell stack. For a discharge time of 1000 s, for example, the heat generation flux in the battery cells is seen in Fig. 7.14 to decline from 3.5 to 2.5 W/m^2 as the battery temperature increases from 15°C to 45°C.

Fig. 7.16 shows the variation of overpotentials within the PEM fuel cell with the current density. It is seen that, as the applied current density increases, the overall cell potential decreases due to the increase in ohmic and activation overpotentials. In addition, the concentration overpotential is seen to be near zero and much smaller than the two other overpotential components. Several points are noted:

- At low current densities, the cell potential is relatively low due to activation overpotential.
- At moderate current densities, the cell potential decreases linearly with current due to ohmic losses.
- At high current densities, the concentration overpotential contributes significantly to the total potential losses within the PEM fuel cell.

Fig. 7.17 shows the variations in the electrical power generation density and energy efficiency of the PEM fuel cell stack with the current density for

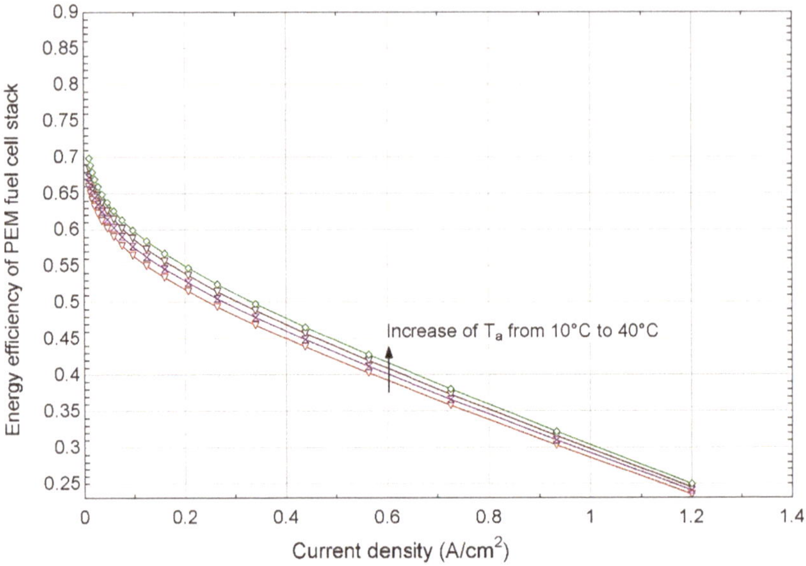

Fig. 7.18 Variation of PEM fuel cell energy efficiency with current density for various temperatures of the air input to the fuel cell.

several air input temperatures. It is seen that, as both temperature and pressure of the input air increase, the electrical power produced from the stack as well as the energy efficiency of the stack increase. The improvement in PEM fuel cell performance with the air temperature can be attributed to the increase in electrolyte conductivity at high temperatures. Higher conductivities lead to lower concentration and ohmic overpotentials because of the higher molecular activity. Therefore, recovering the preheated air used to cool the battery pack instead of directly drawing in air from the environment can improve PEM fuel cell performance. It is also seen in Fig. 7.17 that, at the maximum current density of 1.2 A/cm^2, an increase in the air input temperature from 10°C to 40°C results in an increase in the electrical power density from the fuel cell stack of 0.045 W/cm^2.

In addition, Fig. 7.18 shows the change in the energy efficiency of PEM fuel cell with current density for several air input temperatures. It is observed that the energy efficiency of PEM fuel cell stack improves as the temperature of the input air increases. Also, the PEM fuel cell energy efficiency is higher at relatively lower current densities. At an air input temperature of 40°C, the energy efficiency of the PEM fuel cell decreases from 0.7 to 0.25 as the current density increases from 0 to 1.2 A/cm^2.

7.4.6 Further discussion and closure

In this case study, a novel integration of a battery thermal management system with a PEM fuel cell/lithium-ion battery system is described. The underlying theory of each system is provided along with the corresponding electrochemical reactions. The performance of both the lithium-ion battery cells and the PEM fuel cell stack system is assessed through the overpotential losses, including surface (or activation), concentration and ohmic overpotentials. The lithium-ion battery cell potential and heat generation rates are presented for an isothermal model. These help in the design of a battery thermal management system (in this case based on forced air cooling) and in specifying the temperature range at which the battery can operate effectively. The effect of varying operating temperature is shown to identify how changing this key parameter affects the electrical power production and energy efficiency of the PEM fuel cell stack. It is found that, when using preheated air at a temperature of 40°C, the PEM fuel cell is capable of producing an electrical power density that is 450 W/m^2 more than the nonpreheating condition at cold weather condition (inlet air temperature of 10°C).

It is noted for relatively high ambient temperatures, where the air input temperature to the electrical vehicle is relatively high (above 40°C), that the air cannot effectively cool the battery cells inside the pack. Improper cooling of lithium-ion batteries accelerates their aging mechanisms and consequently lowers the lifetime of the battery cells. It is generally recommended that the operating temperature of lithium-ion battery be maintained between 15°C and 45°C. It is also noted that the effect of cell temperature on the degradation rate for a battery is more severe during charging than discharging.

7.5 Case study 2: Hybrid-electric aircraft propulsion system

There is a growing interest in the use of fuel cells and batteries for electricity production and storage for electric aircraft propulsion systems, as a means of achieving or shifting toward net-zero carbon emissions. The SOFC is used for electrical power generation in aviation applications largely due to its high potential fuel-to-electricity conversion efficiency (Fernandes et al., 2018; Kohout and Schmitz, 2003). But there are numerous issues in such aircraft applications, some of which follow:

- The storage of pure hydrogen (as fuel for the SOFC) in the gas phase requires very large storage tanks, which is a significant disadvantage for mobile applications. So, the hydrogen should be stored as either a compressed gas or a liquid. Storage of hydrogen as a gas typically requires very high pressures (350–700 bar), while the storage of hydrogen as a liquid requires cryogenic temperatures (−252.8°C at atmospheric pressure). However, storage of hydrogen in either liquid or gaseous form can lead to technical difficulties and additional costs.
- The challenges described in the last point regarding the direct use of pure hydrogen in electric aircraft with a SOFC have led to efforts to find alternative hydrogen storage mediums like fuels containing hydrogen. For instance, kerosene, ammonia, methane, ethanol and gasoline are the other fuels can be used in a SOFC. The use of such fuels requires reformation at high temperatures. A reformer is a device that produces hydrogen from a source that contains hydrogen. For short-chain hydrocarbon fuels like methane, natural gas and ethanol, the fuel reformation can take place internally, avoiding the need for an external reformer.
- In addition, a supplemental battery system is often used in the electric vehicle to provide electrical power during takeoff and climbing when the need for power is relatively high. One of the main challenges in the application of batteries in aviation applications is the effect of temperature on battery performance. Therefore, a cooling process for the battery pack in electric aircraft is necessary to maintain a uniform temperature within the desired range for effective and long-term operation of battery cells.

Increasing research is being undertaken by NASA on developing and evaluating for aircraft applications the use of alternative fuels like methane in advanced SOFCs containing internal reformers, and fuel/energy storage technologies (Kohout and Schmitz, 2003). Studies are being carried out to identify concepts with high efficiency and payoff potential and related technologies for further development of electric aircraft. The case study considered here can support that activity.

In this case study, we consider a novel hybrid SOFC-battery-gas turbine propulsion system which supplies electrical power for an electric aircraft. This hybrid system incorporates a lithium-ion battery pack, an SOFC with an internal reformer, compressors, a humidifier, heat exchangers, a gas turbine/generator and other components. System component models are developed to evaluate the performance for the system design, and the components and system are analyzed. Heat generation rates within

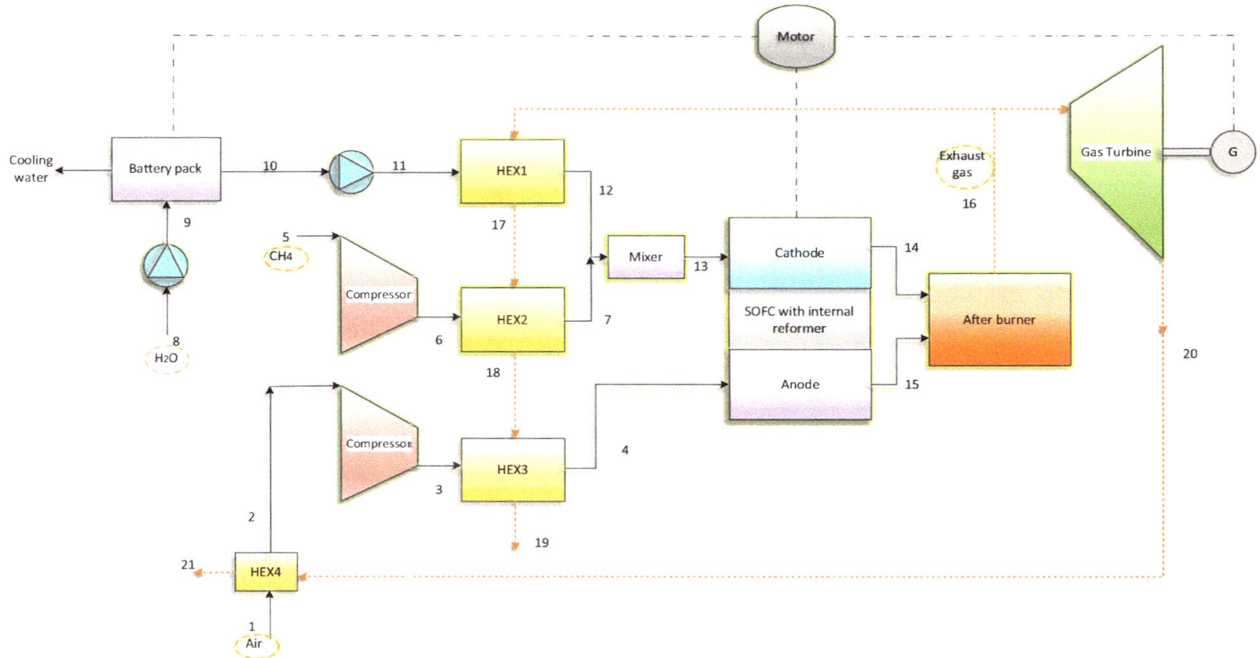

Fig. 7.19 Flow diagram for a hybrid-electric aircraft energy system consisting of a methane-fueled SOFC, a gas turbine and lithium-ion batteries for supplying electrical power.

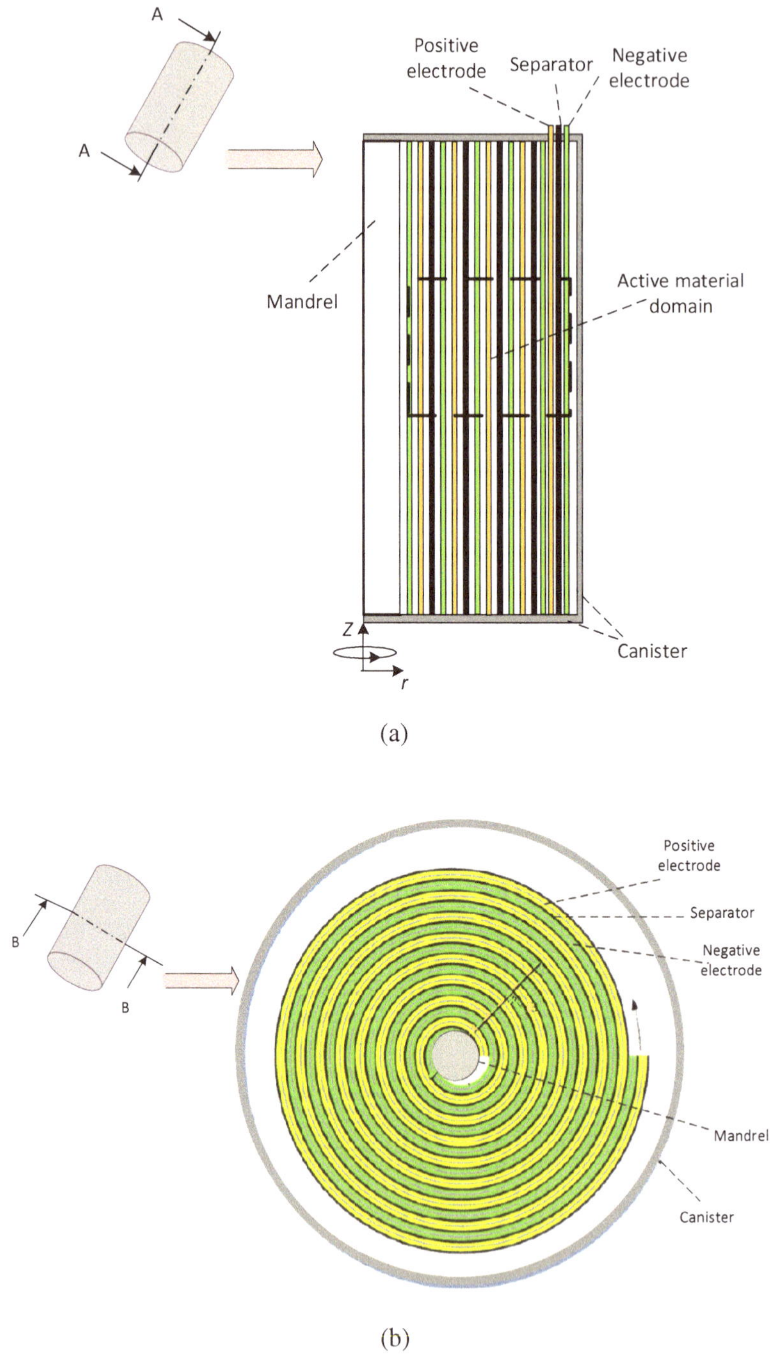

Fig. 7.20 (A) Vertical cross section of half of an upright cylindrical lithium-ion battery. The center of the battery is at the left. (B) Cross section of a cylindrical lithium-ion battery. A spiral roll is composed of anode, cathode and separator. *From Farsi, A., Rosen, M.A., 2023. Performance analysis of a hybrid aircraft propulsion system using solid oxide fuel cell, lithium ion battery and gas turbine. Appl. Energy, 329, 120280.*

lithium-ion battery cells are calculated through via thermal analyses of batteries in which the battery temperature is allowed to change with time. Average values for heat generation rates within lithium-ion battery cells are used in simulations of the remainder of the system components. The electrical power produced by the SOFC with an internal reformer is determined with an electrochemical model and, finally, the performance of overall hybrid system is assessed in terms of its energy efficiency.

7.5.1 System description

In the case study, the design of the hybrid-electric aircraft utilizes a propulsion system in which a SOFC fueled by methane and a gas turbine provide the majority of the electrical power. The lithium-ion batteries are used to supply electrical power during high power load operating conditions.

Fig. 7.19 shows the SOFC-battery-gas turbine system. It includes the following components: air and methane compressors, water pumps, battery pack, heat exchangers, a mixer, a SOFC with an internal reformer, an afterburner, a gas turbine and a motor. The SOFC reactants including water, methane and air are pressurized and preheated in the heat exchangers where their temperature increases to 1000 K.

High battery temperatures shorten its lifetime by increasing the rate of the aging processes within the battery. This makes an effective and efficient cooling necessary for high power applications. In the system considered in the case study, water is pumped to the lithium-ion battery pack to remove part of the heat generated in the battery and to cool the individual cells by passing around them. Excess water exiting the battery pack is discarded from the system while the main portion of water is pumped to the heat exchanger where it is converted to steam and then mixed with high temperature methane in the mixer before entering the SOFC. In the SOFC, which is equipped with an internal reformer, the mixture of steam and methane is first converted to the carbon monoxide and hydrogen through the steam reforming reaction, and then converted to carbon dioxide and hydrogen through the water-gas shift reaction. The produced gases are transferred to the anode side of the SOFC where the hydrogen gas reacts with the oxygen of in the high temperature air at the cathode to produce water.

The unconsumed fuel, excess air and products of the SOFC-internal reformer are utilized in the afterburner for generation of high temperature streams. These are used in the gas turbine for power generation as well as in the heat exchangers for heating the SOFC reactants. The combustion gases exiting the gas turbine can be further utilized for preheating the inlet air to the aircraft. The electrical power produced by the SOFC and gas turbine is used by the aircraft motor, and this usage is assumed constant throughout the flight. Additional power produced during the flight by the gas turbine and SOFC can be stored in the lithium-ion battery and utilized later during takeoff or climbing when much greater electrical power is required.

7.5.2 Modeling and analysis

7.5.2.1 Thermal model of lithium-ion battery

In this case study, the lithium-ion battery is used to assist the SOFC and gas turbine in supplying electrical power during takeoff and climbing. The battery pack is comprised of cylindrical lithium-ion batteries which are cooled through the passage of water between the batteries inside the pack. Fig. 7.20A shows half of a vertical cross section of a cylindrical lithium-ion battery. It contains a mandrel (isolator around which the battery cell sheets are wound), a canister and an active material domain (wound sheets of cell material) including negative and positive electrodes and a separator. The height of the battery cell is 65 mm and the radius is 9 mm. The thickness of the steel canister in 0.25 mm and the radius of the mandrel is 2 mm. For a cylindrical battery, as shown in Fig. 7.20, the separator, anode and cathode sheets are cut into two long strips which are wound on a cylindrical mandrel to form a cylindrical structure.

A detailed electrochemical model for a lithium-ion battery was presented previously for case study 1 (see Section 7.4.1). That model is used as the basis of the analysis in this case study, including determination of the heat generated in the lithium-ion battery. In case study 1, the electrochemical equations are applied for a one-dimensional model of a battery cell, while in the present case study those equations are applied in a two-dimensional and axially symmetric model of the battery cell. The reason for considering an axial-symmetric condition is that heat conduction in the spiral direction (direction through which the anode, cathode and separator layers are rolled around the mandrel) can be neglected for a spirally wound battery cell (see Fig. 7.20B).

In case study 1, the battery cell was assumed to have a uniform and constant temperature during charging and discharging. In this case study, however, the battery temperature is allowed to vary with time (i.e., we have a nonisothermal condition). Also, since the thermal conductivities of the components of the lithium-ion battery cell are relatively high and the heat generated in the cell is not excessive, it can be assumed that the battery has a uniform temperature profile. In addition, when the chemistry of the battery is not highly affected by small changes in the battery temperature, using a lumped parameter approach which is based on the average temperature can predict the battery thermal behavior with adequate accuracy. In the rest of this section, the thermal model of a lithium-ion battery cell is presented and applied to determine the battery temperature variations over time. The developed electrochemical and thermal models are coupled through a multiphysics coupling node in the Comsol software to connect the heat generated in the battery and the temperature of the battery.

The general energy equation for a multicomponent system (like a battery), which is derived from a differential energy balance, can be written as:

$$\frac{\partial(\rho_B c_{p,B} T_B)}{\partial t} = \nabla .k\nabla T + \dot{Q}_{gen} \tag{7.54}$$

where $c_{p,\ B}$, ρ_B, T_B, and t respectively denote specific heat at constant pressure, density of battery, battery temperature and time. Also, $\dot{Q}_{gen}$ is the heat generation rate per unit volume for the battery.

For a two-dimensional model of a cylindrical battery cell considered here, the above equation can be written as

$$\frac{\partial(\rho_B c_{p,B} T_B)}{\partial t} = \frac{1}{r}\frac{\partial}{\partial r}\left(k_r r\frac{\partial T_B}{\partial r}\right) + \frac{\partial}{\partial z}\left(k_z\frac{\partial T_B}{\partial z}\right) + \dot{Q}_{gen} \tag{7.55}$$

where k_r and k_z respectively are the thermal conductivities of the cell in the radial and axial directions. For a single cell, which has a small thickness, the *lumped-parameter approach* is applicable. This indicates the condition where the ratio of convective heat transfer coefficient (h_{conv}) times the thickness of the cell (δ_B) to the thermal conductivity of the cell (k) is very small (i.e., $h_{conv}\delta_B/k \ll 1$). Hence, the assumption of uniform temperature throughout the cell is reasonably accurate.

The heat generation rate in the battery is due to heat of electrochemical reactions (reversible heat) and internal resistances (irreversible heat) within the cell. A general expression for the battery heat generation rate per volume of the cell can be written as:

$$\dot{Q}_{gen} = \frac{1}{V_B}\left(I_B(E_{OCP} - E_B) - I\left(T_B\frac{\partial E_{eq}}{\partial T}\right)\right) \tag{7.56}$$

where E_{eq}, I_B, E_B, and V_B respectively denote equilibrium cell potential, cell current, cell voltage and volume of cell. The reversible heat rate term in Eq. (7.56) (i.e., $I(T_B\partial E_{eq}/\partial T_B)$) is equivalent to $T_B\Delta SI_B/n_eF$ in which ΔS is the entropy change due to electrochemical reactions occurring in the cell and n_e is the number of electrons in the balanced electrochemical half reaction.

Each battery is comprised of three layers (anode, cathode, and separator) which have different thermal conductivities in different directions of the cylindrical battery cell. The battery overall thermal conductivities in radial (k_r) and axial (k_z) directions respectively can be expressed as:

$$k_r = \frac{\sum \delta_i}{\sum (\delta_i / k_i)} \tag{7.57}$$

$$k_z = \frac{\sum \delta_i k_i}{\sum (\delta_i)} \tag{7.58}$$

where δ_i and k_i are the thickness of layer i and thermal conductivity of layer i. Different values for these parameters apply to the different materials constituting the layers, respectively.

Similarly, the density and specific heat at constant pressure of the battery with its multilayer structure can be written respectively as:

$$\rho_B = \frac{\sum \delta_i \rho_i}{\sum (L_i)} \tag{7.59}$$

$$c_{p,B} = \frac{\sum \delta_i c_{p,i}}{\sum (\delta_i)} \tag{7.60}$$

where ρ_i and $c_{p,\,i}$ are the density and specific heat capacity of layer i in the battery cell. The water that is used for the SOFC first is passed through the lithium-ion battery pack for cooling purposes and then is pumped through the heat exchanger. To account for the cooling effect of water between the battery cells, a heat flux boundary condition is specified on the battery canister surface. Doing so allows the convective heat transfer between the cooling water and the battery surface to be evaluated. The input parameters for the transient model for the lithium-ion battery are listed in Table 7.3, along with their values.

Table 7.3 Input design parameters for transient model of the lithium-ion battery cell.

Parameter	Value
Thickness of battery canister	2.5×10^{-4} m
Battery radius	9×10^{-3} m
Battery height	6.5×10^{-2} m
Mandrel radius	2×10^{-3} m
Length of negative electrode	5.5×10^{-5} m
Length of separator	3×10^{-5} m
Length of positive electrode	5.5×10^{-5} m
Thickness of negative current collector	7×10^{-6} m
Thickness of positive current collector	1×10^{-5} m
Cell thickness	1.57×10^{-4} m
Thermal conductivity of positive electrode	1.58 W/(m K)
Thermal conductivity of negative electrode	1.04 W/(m K)
Thermal conductivity of positive current collector	170 W/(m K)
Thermal conductivity of negative current collector	398 W/(m K)
Thermal conductivity of separator	0.344 W/(m K)
Density of positive electrode	2328.5 kg/m^3
Density of negative electrode	1347.3 kg/m^3
Density of positive current collector	2770 kg/m^3
Density of negative current collector	8933 kg/m^3
Density of separator	1009 kg/m^3
Heat capacity of positive electrode	1269.2 J/(kg K)
Heat capacity of negative electrode	1437.4 J/(kg K)
Heat capacity of positive current collector	875 J/(kg K)
Heat capacity of negative current collector	385 J/(kg K)
Heat capacity of separator	1978.2 J/(kg K)
Battery angular thermal conductivity	29.557 W/(m K)
Battery radial thermal conductivity	0.89724 W/(m K)
Battery density	2055.2 kg/m^3
Battery heat capacity	1399.1 J/(kg K)
Inlet temperature	298.15 K
Initial temperature	298.15 K
Cell capacity	14,400C
Ohmic overpotential at 1C	0.0045 V
C-rate	2

Data from Chen, S.C., Wang, Y.Y., Wan, C.C., 2006. Thermal analysis of spirally wound lithium batteries. J. Electrochem. Soc. 153(4), A637. Gomadam, P.M., White, R.E., Weidner, J.W., 2003. Modeling heat conduction in spiral geometries. J. Electrochem. Soc. 150(10), A1339. Chen, S.C., Wan, C.C., Wang, Y.Y., 2005. Thermal analysis of lithium-ion batteries. J. Power Sources 140(1), 111–124.

7.5.2.2 Solid oxide fuel cell model

The SOFC is a high-temperature (600–1000°C), all-ceramic and solid-state unit in which an oxide ion conducting ceramic material is used as the electrolyte. A SOFC has potential advantages over PEM fuel cells for aircraft applications. For instance, the relatively higher operating temperature of a SOFC results in greater opportunities to extract waste heat from the SOFC product streams that can be used for other processes in the system such as fuel heating and reformation.

The model and analysis of the SOFC is carried out here under equilibrium and steady state conditions. The following assumptions are invoked, following (Ranjbar et al., 2014; Farsi and Rosen, 2023), for the methane-fueled SOFC equipped with an internal reformer:

- Air is composed of 21% oxygen and 79% nitrogen on a volume basis.
- The temperature difference between the SOFC reactants and products is fixed at 50 K.
- Radiation heat transfer between the solid structures and gas channels is negligible.
- Unused fuel that is not consumed in the SOFC is taken to be oxidized completely in the afterburner.
- The SOFC is insulated fully, so there is no heat interaction with the environment.
- Contact resistances can be neglected.
- Only H_2 participates in the electrochemical reaction with oxygen, and carbon monoxide is converted to carbon dioxide and hydrogen through the water–gas shift reaction.

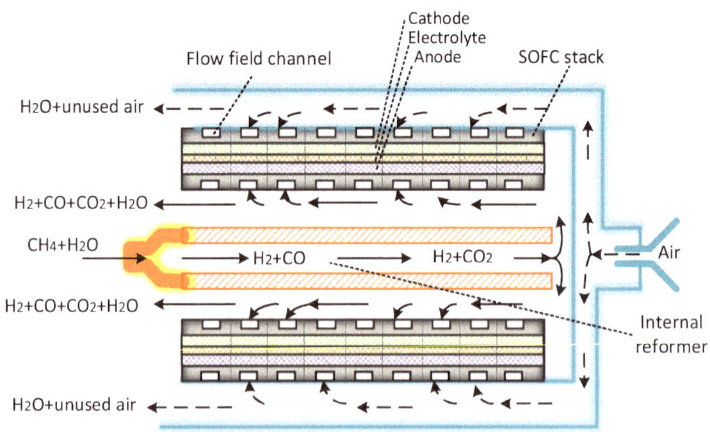

Fig. 7.21 Solid oxide fuel cell with internal reformer.

- Chemical equilibrium is taken to exist for the gas mixture at the exit of the fuel channel.

For a methane-fueled SOFC, it is more economical to use an internal reformer rather than an external reformer. Fig. 7.21 shows an internal reformer combined with a SOFC. In the reforming of methane with steam, the main chemical reactions are as follows:

$$CH_4 + H_2O \rightarrow CO + 3H_2 \tag{7.61}$$

$$CO + H_2O \rightarrow CO_2 + H_2 \tag{7.62}$$

The steam reforming reaction in Eq. (7.61) is slow and endothermic while the water-gas shift reaction in Eq. (7.62) is fast and exothermic. The heat interaction required for steam reforming is much higher than for the water-gas shift reaction (i.e., at a reaction temperature of 1300 K, the steam reforming enthalpy of reaction is 227.5 kJ/mol and for the water-gas shift reaction is −31.8 kJ/mol). So, the overall reaction in the internal reformer is endothermic. The hydrogen produced in the steam reforming and water-gas shift reactions is used for the electrochemical reaction in the SOFC. The overall reaction at the SOFC (i.e., $H_2 + O_2 \rightarrow H_2O$) is exothermic. So, the generated heat by the SOFC can be used to supply heat for steam reforming in the reformer. This reduces the dependency of the internal reformer-SOFC system on a cooling system. The temperature difference between the reactants at the SOFC inlet and the products at the SOFC outlet is assumed here to be 50 K. In addition, in the internal reformer-SOFC system, the unused fuel can be utilized in the afterburner to produce heat for the heat exchangers that heat the SOFC reactants, as well as in the gas turbine for electrical power production.

In the remainder of this subsection, a zero-dimensional model is developed for SOFC which is assumed to operate at steady state conditions. Like the PEM fuel cell that is modeled in case study 1, the output voltage of the SOFC can be expressed as:

$$E_{SOFC} = E_{OCP,SOFC} - \left(\left(\eta_{a,a} + \eta_{a,c}\right) + \left(\eta_{c,a} + \eta_{c,c}\right) + \eta_{ohm}\right) \tag{7.63}$$

where $E_{OCP,\ SOFC}$ is the open circuit potential of the SOFC, $\eta_{a,\ a}$ and $\eta_{a,\ c}$ are the activation overpotentials at the anode and cathode respectively, $\eta_{c,\ a}$ and $\eta_{c,\ c}$ are the concentration overpotentials at the anode and cathode respectively and η_{ohm} is the ohmic overpotential. The calculations of the equilibrium potential and the voltage losses in the SOFC are presented and described in Table 7.4.

Table 7.4 Equations for determination of equilibrium potential and overpotentials in the SOFC.

Potential type	Equation	Description
Equilibrium potential, $E_{OCP,\ SOFC}$	$E_{OCP,SOFC} = -\frac{\Delta\bar{g}^{\circ}}{2F} + \frac{RT_{SOFC}}{2F}\ln\left(\frac{a_{H_2,14}a_{O_2,15}{}^{0.5}}{a_{H_2O,14}}\right)$ $a_{H_2O,14} = \frac{P_{H_2O,14}}{P_{ref}} = \frac{x_{H_2O,14}P_{14}}{P_{ref}}$ $a_{H_2,14} = \frac{P_{H_2,14}}{P_{ref}} = \frac{x_{H_2,14}P_{14}}{P_{ref}}$ $a_{O_2,15} = \frac{P_{O_2,15}}{P_{ref}} = \frac{x_{O_2,15}P_{15}}{P_{ref}}$ $\Delta\bar{g}^{\circ} = \bar{g}^{\circ}_{H_2O} - \bar{g}^{\circ}_{H_2} - 0.5\bar{g}^{\circ}_{O_2}$ $\bar{g}^{\circ} = \bar{h} - T_{SOFC}\bar{s}^{\circ}$	Parameters a_i and x_i respectively are the activity factor of species i in the electrochemical reaction and the molar concentration of species i. Also, $\bar{g}^{\circ}$, $\bar{h}$, and $\bar{s}^{\circ}$ denote molar Gibbs free energy, molar enthalpy and molar entropy, respectively. P_i denotes partial pressure of species i
Activation overpotential, η_a	$\eta_a = \eta_{a,\ a} + \eta_{a,\ c}$ $\eta_{a,a} = \frac{RT_{SOFC}}{F}\left(\sinh^{-1}\left(\frac{i}{2i_{oa}}\right)\right)$ $\eta_{a,c} = \frac{RT_{SOFC}}{F}\left(\sinh^{-1}\left(\frac{i}{2i_{oc}}\right)\right)$	Parameters i, i_{oa}, and i_{oa} denote current density, exchange current density of anode and exchange current density of cathode, respectively. T_{SOFC} and R respectively are SOFC temperature and universal gas constant
Ohmic overpotential, η_{ohm}	$\eta_{ohm} = \left(R_c + \sum_j \rho_j\delta_j\right)i$ $\rho_{int} = 1/(9.3\times10^6\exp(-1100/T_{SOFC}))$ $\rho_c = 1/(42\times10^4\exp(-1200/T_{SOFC}))$ $\rho_e = 1/(3.34\times10^4\exp(-10300/T_{SOFC}))$ $\rho_a = 1/(95\times10^6\exp(-1150/T_{SOFC}))$	Parameters ρ_{int}, ρ_c, ρ_e, and ρ_a denote electrical resistivity of interconnect, cathode, electrolyte and anode, respectively. Also, δ_i is the thickness of layer i of the SOFC
Concentration overpotential, η_c	$\eta_c = \eta_{c,\ a} + \eta_{c,\ c}$ $\eta_{c,a} = \frac{RT_{SOFC}}{2F}\left(\ln\left(1 + \frac{P_{H_2}i}{2P_{H_2O}i_{as}}\right) - \ln\left(1 - \frac{i}{i_{as}}\right)\right)$ $\eta_{c,c} = -\frac{RT_{SOFC}}{4F}\ln\left(1 - \frac{i}{i_{cs}}\right)$ $j_{as} = 2FP_{H_2}D_{a,\ eff}/(RT_{SOFC}\delta_a)$ $j_{ac} = 4FP_{O_2}D_{c,eff}/\left(\frac{P_{15}-P_{O_2,15}}{P_{15}}(RT_{SOFC}\delta_c)\right)$	Parameters $\eta_{c,\ a}$ and $\eta_{c,\ c}$ respectively are concentration overpotentials at anode and cathode Parameters i_{as} and i_{cs} respectively are anode and cathode limiting current densities. Also, $D_{a,\ eff}$ and $D_{c,\ eff}$ are effective gaseous diffusivities through the anode and cathode respectively, while δ_a and δ_c are thicknesses of the anode and cathode respectively

Sources: Mozdzierz, M., Chalusiak, M., Kimijima, S., Szmyd, J.S., Brus, G., 2018. An afterburner-powered methane/steam reformer for a solid oxide fuel cells application. Heat Mass Transf. 54(8), 2331–2341. Yahya, A., Ferrero, D., Dhahri, H., Leone, P., Slimi, K., Santarelli, M., 2018. Electrochemical performance of solid oxide fuel cell: experimental study and calibrated model. Energy 142, 932–943.

At the equilibrium condition, the molar conversion rates for the reforming and shift chemical reactions in the internal reformer are denoted r_1 and r_2, while the molar conversion rate for the overall electrochemical reaction in the SOFC is denoted r_3. These rates can be evaluated using the equilibrium constant and current relations. The equilibrium constant for the shift reaction can be written as

$$\ln K_s = -\frac{\Delta \overline{g}_s^{\circ}}{RT_{SOFC}} = \ln\left[\frac{r_2(3r_1 + r_2 - r_3)}{(r_1 - r_2)(1.5r_1 - r_2 + r_3)}\right] \tag{7.64}$$

where K_s is the equilibrium constant for the reaction. The specific Gibbs free energy can be given as:

$$\Delta \overline{g}_s^{\circ} = \overline{g}_{s,H_2}^{\circ} + \overline{g}_{s,CO_2}^{\circ} - \overline{g}_{s,H_2O}^{\circ} - \overline{g}_{s,CO}^{\circ} \tag{7.65}$$

Here the specific Gibbs free energy condition can be determined as $\overline{g}_s^{\circ} = \overline{h} - T_{SOFC}\overline{s}^{\circ}$ where T_{SOFC} is the temperature of the SOFC products. The molar enthalpy ($\overline{h}$) is evaluated at T_{SOFC} and the molar entropy ($\overline{s}^{\circ}$) at T_{SOFC} and standard pressure (101 kPa). Based on the Faraday law, the molar consumption rate of methane at the anode can be written as

$$\dot{N}_{CH_4,consumed} = \frac{i_{SOFC}.N_{SOFC}.A_{SOFC}}{2F} \tag{7.66}$$

where i_{SOFC} is the net exchange current density of the SOFC, N_{SOFC} the number of fuel cells in the stack, A_{SOFC} the active area of the cell and F the Faraday constant.

The fuel utilization factor can be expressed as:

$$U_F = \frac{\dot{n}_{CH_4,consumed}}{\dot{n}_{CH_4,in}} \tag{7.67}$$

where $\dot{n}_{CH_4}$ is molar flow rate of methane. The molar flow rate of working fluids in the SOFC system with the internal reformer can be calculated through a mass balance. Table 7.5 lists the mass balances for the streams of the examined system. In addition, the input design parameters used to simulate the SOFC with the internal reformer are presented in Table 7.6.

7.5.2.3 Energy analysis of the system

A general energy rate balance for components of the system under steady state conditions and neglecting and potential energy values can be written as (Farsi and Rosen, 2022a):

Table 7.5 Molar flow rate equations for the SOFC with internal reformer.

Description	Equations
Chemical reactions in the SOFC with internal reformer and the corresponding molar conversion ratio	$r_1[CH_4 + H_2O \rightarrow CO + 3H_2]$ $r_2[CO + H_2O \rightarrow CO_2 + H_2]$ $r_3[H_2 + 0.5O_2 \rightarrow H_2O]$
Mass balance in the SOFC with internal reformer	$\dot{n}_{CH_4,13} = r_1$ $\dot{n}_{H_2O,13} = x_{sc}\dot{n}_{CH_4,13} = 2.5r_1$ $x_{sc} = \dot{n}_{H_2O,13}/\dot{n}_{CH_4,13}$ $\dot{n}_{H_2,14} = 3r_1 + r_2 - r_3$ $\dot{n}_{CO,14} = r_1 - r_2$ $\dot{n}_{CO,14} = r_2$ $\dot{n}_{H_2O,14} = \dot{n}_{H_2O,13} - r_1 - r_2 + r_3 = 1.5r_1 - r_2 + r_3$ $\dot{n}_{O_2,4} = r_3/2U_a$ air utilization ratio $(U_a) = \frac{consumed\ O_2}{supplied\ O_2} = \frac{(r_3/2)}{\dot{n}_{O_2,4}}$ $\dot{n}_{N_2,15} = \dot{n}_{N_2,4}$ $\dot{n}_4 = \dot{n}_{O_2,4} + \dot{n}_{N_2,4}$ $\dot{n}_{15} = \dot{n}_{O_2,15} + \dot{n}_{N_2,15}$ $\dot{n}_{14} = \dot{n}_{H_2,14} + \dot{n}_{H_2O,14} + \dot{n}_{CO,14} + \dot{n}_{CO_2,14}$ $\dot{n}_{13} = \dot{n}_{H_2O,13} + \dot{n}_{CH_4,13}$

Table 7.6 Input design parameters considered for the SOFC/internal reformer model.

Parameter	Value
U_F	85%
A_{SOFC}	$100\,cm^2$
i_{SOFC}	$0.8\,A/cm^2$
i_{oa}	$0.56\,A/cm^2$
i_{oc}	$0.25\,A/cm^2$
$D_{a,\ eff}$	$0.2 \times 10^{-4}\ m^2/s$
$D_{c,\ eff}$	$0.5 \times 10^{-5}\ m^2/s$
δ_a	0.5×10^{-3} m
δ_c	0.5×10^{-4} m
δ_e	0.1×10^{-4} m
δ_{int}	0.3×10^{-2} m
x_{sc} (steam to carbon ratio)	2.5
SOFC pressure drop	2%
Number of cells in a stack	100
Number of stacks	34

Data from Colpan, C.O., Dincer, I., Hamdullahpur, F., 2007. Thermodynamic modeling of direct internal reforming solid oxide fuel cells operating with syngas. Int. J. Hydrogen Energy 32(7), 787–795. Yahya, A., Ferrero, D., Dhahri, H., Leone, P., Slimi, K., Santarelli, M., 2018. Electrochemical performance of solid oxide fuel cell: experimental study and calibrated model. Energy 142, 932–943.

$$\dot{Q} - \dot{W} + \sum_{in} \dot{n}_i \overline{h}_i - \sum_{out} \dot{n}_o \overline{h}_o = 0 \tag{7.68}$$

where $\dot{n}$ and $\overline{h}$ respectively denote mole flow rate and molar enthalpy, and $\dot{Q}$ and $\dot{W}$ respectively denote heat transfer rate and work rate. The isentropic efficiencies for the air/methane compressors and water pump are all assumed to be 85%. The pressure drop in the afterburner and heat exchangers are assumed to be 3% and 2% respectively of the corresponding input pressures (Ranjbar et al., 2014).

The ambient temperature and pressure vary according to altitude (Z), with both decreasing with increasing altitude. The ambient temperature and ambient pressure can be determined as follows:

$$T_a = 288.15 + L_a Z \tag{7.69}$$

$$P_a = 101.325 \left(\frac{288.15}{T_a} \right)^{\frac{g}{L_a R_a}} \tag{7.70}$$

where R_a, g, and L_a are the characteristic gas constant for air, gravitational acceleration and the base temperature lapse rate per kilometer of geopotential altitude ($L_a = 6.5\,\text{K/km}$) (Seyam et al., 2021). The inlet air condition to the system is calculated as:

$$P_{air,in} = P_a \left[1 + \frac{(\gamma - 1)}{2} M^2 \right]^{(\gamma - 1)^{-1}} \tag{7.71}$$

$$T_{air,in} = T_a \left[1 + \frac{(\gamma - 1)}{2} M^2 \right] \tag{7.72}$$

where γ and M respectively are the specific heat ratio of air (1.4) and Mach number. The Mach number for an altitude of 10 km is 0.83 (Seyam et al., 2021).

It is assumed that 40% of the combusted flue gas leaving the afterburner is transferred to the gas turbine for electrical power generation and that the remaining portion (60%) is conveyed to the heat exchangers for preheating of SOFC reactants. The gas turbine outlet temperature is determined as follows:

$$T_{20} = T_{16} \left[1 - \eta_{isen,T} \left[1 - \left(\frac{P_{16}}{P_{20}} \right) \right] \right] \tag{7.73}$$

where $\eta_{isen,\ T}$ is the isentropic efficiency of the gas turbine, considered here to be 88%. The net electrical power output of the SOFC and gas turbine can be obtained as the difference between the total electrical power produced by the gas turbine and SOFC and the total required power for pumps and compressors (and any other devices requiring electricity). That is,

$$\dot{W}_{net} = \dot{W}_{SOFC} + \dot{W}_{GT} - \sum_{compressors} \dot{W}_i + \sum_{pumps} \dot{W}_j \tag{7.74}$$

The net fuel-to-electricity conversion efficiency compares net power produced by the SOFC-battery-gas turbine propulsion system to the input fuel consumption, as follows:

$$\eta_{SOFC-GT} = \frac{\dot{W}_{net}}{\dot{n}_{CH_4,in}\overline{LHV}_{CH_4}} \tag{7.75}$$

where $\overline{LHV}_{CH_4}$ is low heating value of methane and $\dot{n}_{CH_4,in}$ is the molar flow rate of methane to the hybrid propulsion system.

7.5.3 Results and discussion

The results of the modeling and analysis of the SOFC-battery hybridized gas turbine propulsion system for electric aircraft are presented and discussed in this section. The aircraft is assumed to be in cruising condition at an altitude of 10 km, where the ambient conditions are 254 K and 36.45 kPa.

More specifically, the effects are assessed and reported of varying the main system parameters, like SOFC temperature and pressure and current density, on the net electrical power produced and the energy efficiency for the hybrid SOFC-battery-gas turbine.

In addition, the results of a thermal study of the lithium-ion battery are presented. The thermal model of the lithium-ion battery is simulated in Comsol multiphysics software. The average values for the temperature and heat generation rate in the lithium-ion battery are transferred to EES, where the SOFC stack, afterburner, gas turbine and compressors are modeled. The simulation results presented in this section for the lithium-ion battery are for a discharge rate of 16 A, an initial temperature of 298 K and an initial state of charge of 20%. The charge and discharge durations are taken to be 600 s each.

In the remainder of this section, the results obtained from applying the model for the lithium-ion battery and SOFC are described.

Fig. 7.22 shows a single charge-discharge cycle for the lithium-ion battery. The charge and discharge durations are 600 s each. A positive current for the battery is associated with charging and a negative sign with discharging.

Fig. 7.23 compares the battery open circuit potential and cell potential, as they vary with time. It is seen that the battery open circuit potential is higher than the cell potential during discharge and is lower during charge because during the charge and discharge operations of the battery, the cell potential experiences ohmic, surface and concentration overpotentials.

Fig. 7.24 shows the variation of battery maximum temperature during discharging and charging periods for several C-rates. It is seen that the battery temperature increases consistently over the operation of battery. In addition, higher temperatures are observed at higher C-rates. For instance, at a time of 500s, the temperature of the lithium-ion battery increases from 27°C to 44°C as the C-rate increases from 1 to 4. This graph is particularly useful as it provides the maximum allowable temperature of the battery that can be tolerated to prevent melting of solid parts of the electrodes or thermal decomposition of the polymer in the lithium-ion battery.

Fig. 7.25 shows the variations with time of the heat generation rate per volume (i.e., volumetric heat generation rate) for the lithium-ion battery

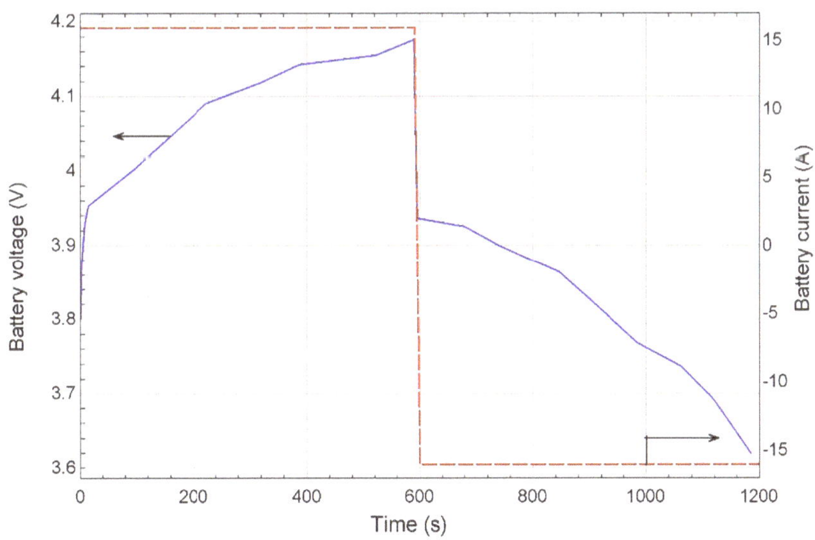

Fig. 7.22 Variation with time of lithium-ion battery cell potential and the load cycle current.

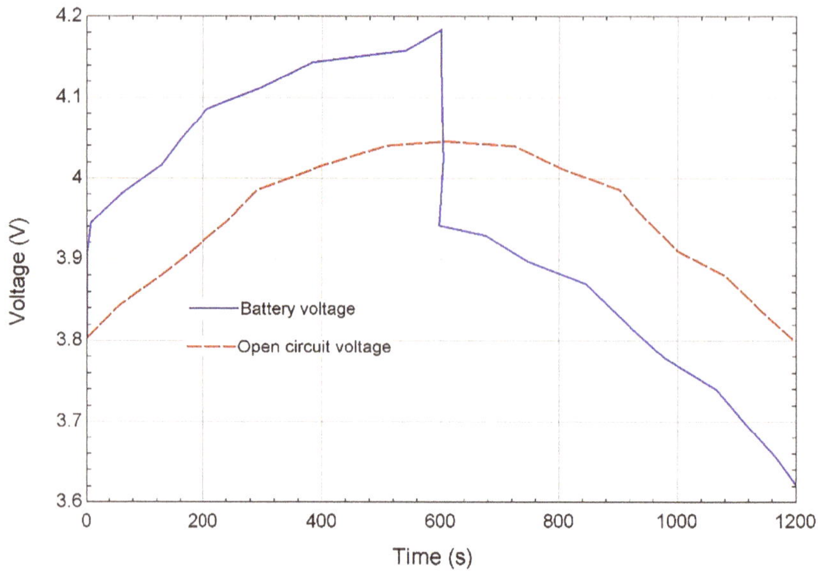

Fig. 7.23 Cell potential and open circuit potential variations with time during operation of the lithium-ion battery cell.

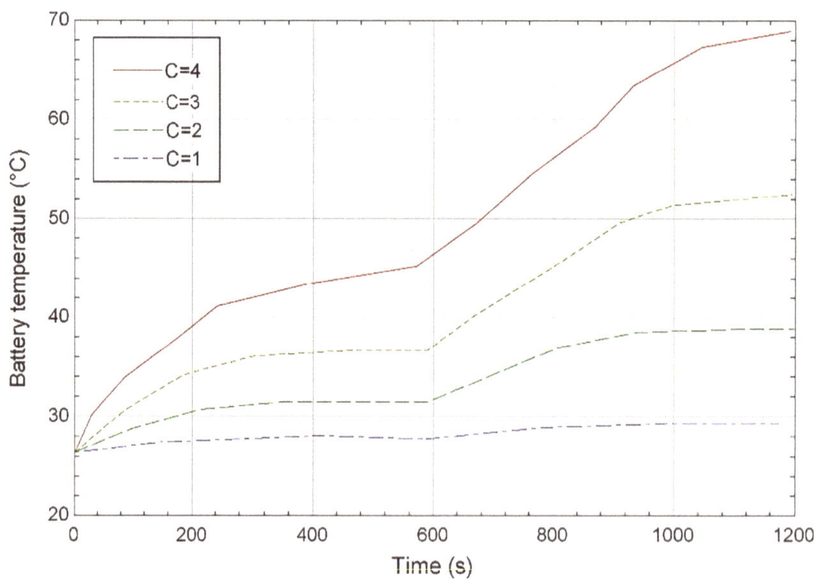

Fig. 7.24 Variation of battery maximum temperature with charging and discharging time for several C-rates.

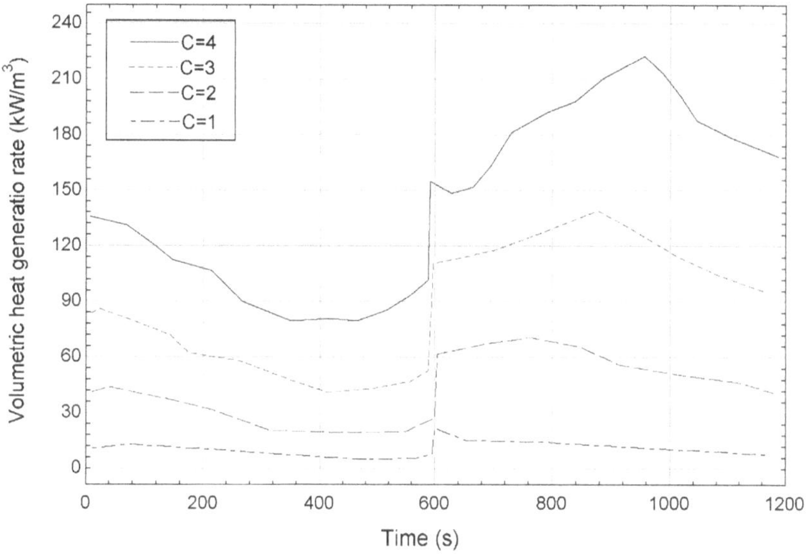

Fig. 7.25 Variation of battery volumetric heat generation with charging and discharging time for several C-rates.

during charging and discharging periods, for several C-rates. According to Fig. 7.25, during charging, the volumetric heat generation rate decreases because the increase in the battery temperature (Fig. 7.25) reduces the overpotentials, which constitute the primary ohmic loss. Then near the end of charging, the volumetric heat generation rate increases as the further increase in battery temperature (Fig. 7.25) leads to an increase in the overpotentials. Also, during discharging, the volumetric heat generation rate in the battery increases due to the increase in the overpotentials (mainly ohmic overpotential) and then this heat rate decreases due to decrease in the overpotentials (mainly concentration overpotential). Note that the battery temperature increases consistently during both charge and discharge (Fig. 7.24).

Fig. 7.26 illustrates the temperature distribution through a single lithium-ion battery cell selected times. It is seen that the temperature of various parts of battery does not change significantly. For instance, at the end of discharge (1119 s), the difference between battery maximum and minimum temperatures is 4°C. This demonstrates the validity of the assumption of uniform temperature throughout the lithium-ion battery cell.

Fig. 7.27 shows the variations with the current density in the SOFC of the electrical power produced by the SOFC and gas turbine as well as the

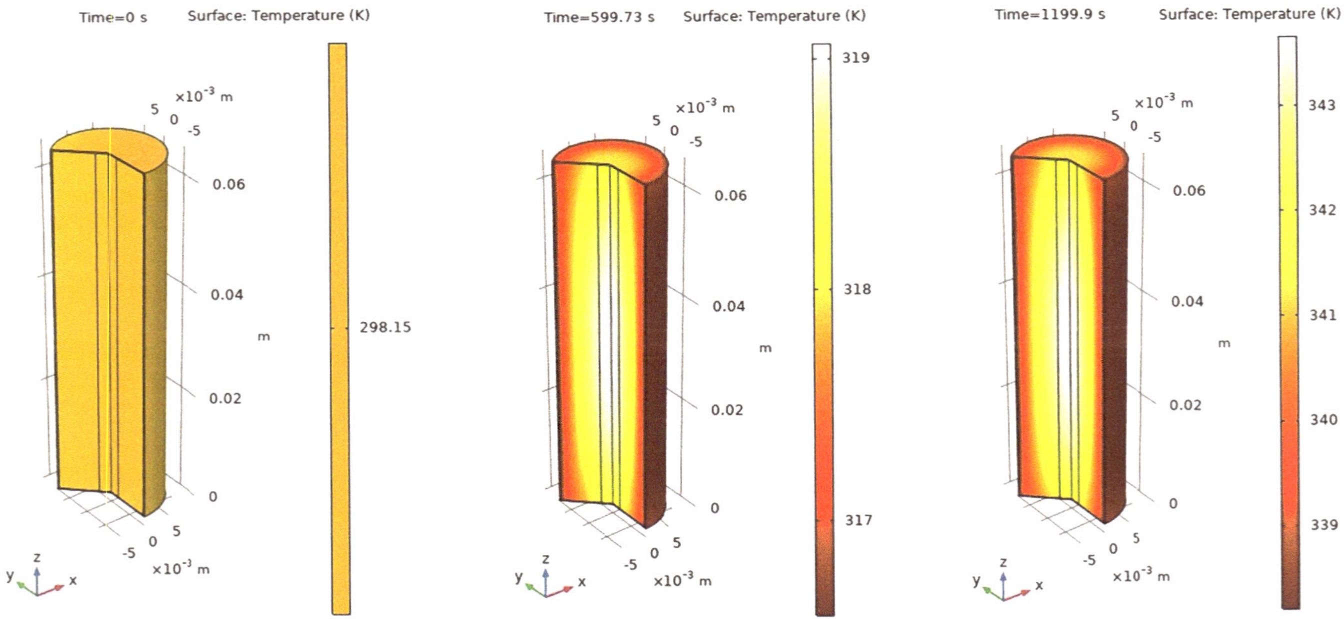

Fig. 7.26 Temperature distribution in the lithium-ion battery cell at beginning of charging (left), end of charging (center) and end of discharging (right).

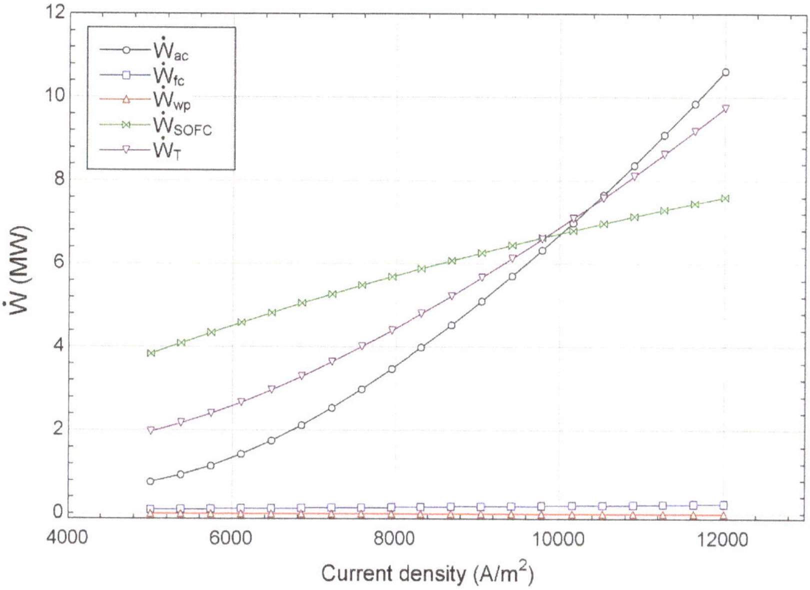

Fig. 7.27 Variations with the current density in the SOFC of the electrical power production/consumption components for the hybrid SOFC-battery-gas turbine system.

power consumed by the air compressor, fuel compressor and water pumps. It is seen that the slope of the increase is lower for the SOFC power production than the gas turbine and air compressor. The reason is that increasing the current density raises the overpotentials, primary activation and ohmic losses in the SOFC, reducing the fuel cell potential. But the decline in the cell potential is smaller than the rise in the current density, which results in an increase in the electrical power output by the SOFC. It is also seen that the increase in the current density of the SOFC leads to an increase in all of the electrical power producer/consumer components. Higher current densities result in higher flow rates of reactants (air, methane, and water) and consequently more power is required to pressurize those reactants to the SOFC pressure. The electrical power consumed for the fuel compressor and water pump is lower than that for the air compressor, as the input rate is much higher for air than for methane and water.

Fig. 7.28 shows the variations with SOFC current density of total net electrical power yielded by the hybrid SOFC-battery-gas turbine system and its energy efficiency in the electric aircraft. It is observed that the net electrical power produced by the hybrid system is a maximum (6.7 MW) at a current density of 10,000 A/m^2. However, the energy efficiency of

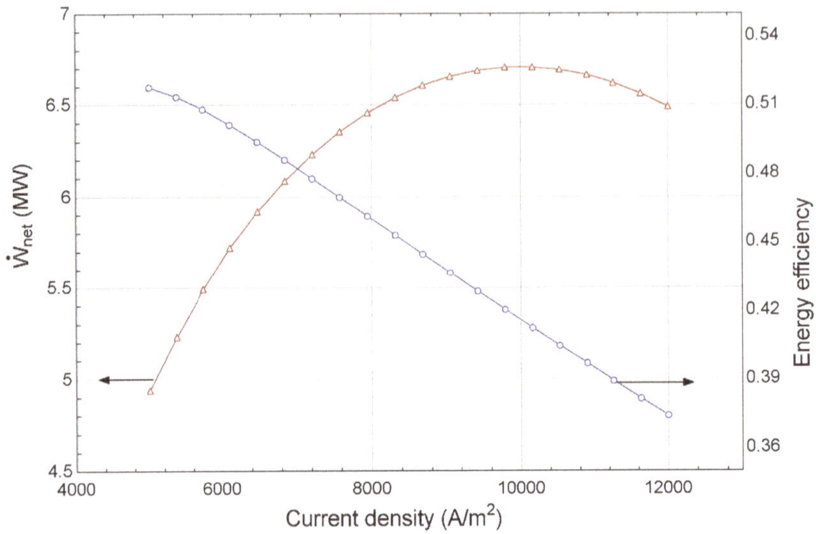

Fig. 7.28 Variations of net electrical power produced by and energy efficiency of the hybrid SOFC-battery-gas turbine system with SOFC current density.

the hybrid system decreases continually with increasing current density. This is mainly due to the fact that the increase in the input flow rate of methane at higher current densities is greater than the increase in the net electrical power output from the hybrid system. Therefore, it is suggested that the current density of SOFC not exceed 10,000 A/m^2.

Fig. 7.29 shows the effect of pressure ratio on the net electrical power produced by and energy efficiency of the hybrid system. The pressure ratio indicates the ratio of output to input pressure of the SOFC reactants, and is applicable to the compressors and pump. As stated earlier, the SOFC reactants of water, methane and air are pressurized to the SOFC pressure before entering the internal reformer. A higher pressurization ratio results in higher electrical power production by the SOFC and gas turbine, as well as higher power consumption by the air/fuel compressors and water pump. As a consequence, the net power produced from the system reaches maximum at a specific pressurization ratio. The energy efficiency of the hybrid system is also a maximum at that pressurization ratio. From Fig. 7.29 it is seen that the maximum net power output and energy efficiency are achieved at pressurization ratio of approximately 7.5.

Fig. 7.30 shows the variation of electrical power produced or consumed by the hybrid system components, with the temperature of SOFC. It can be

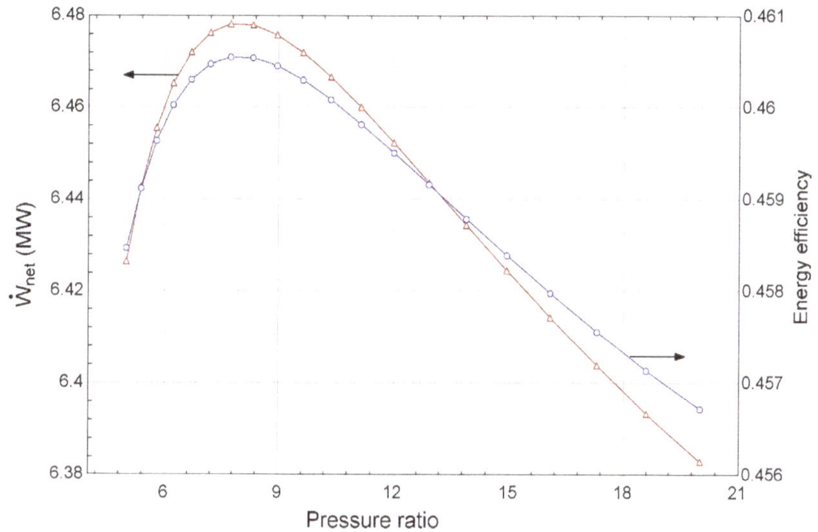

Fig. 7.29 Variation with pressure ratio of SOFC reactants of net electrical power produced and energy efficiency for the hybrid SOFC-battery-gas turbine system.

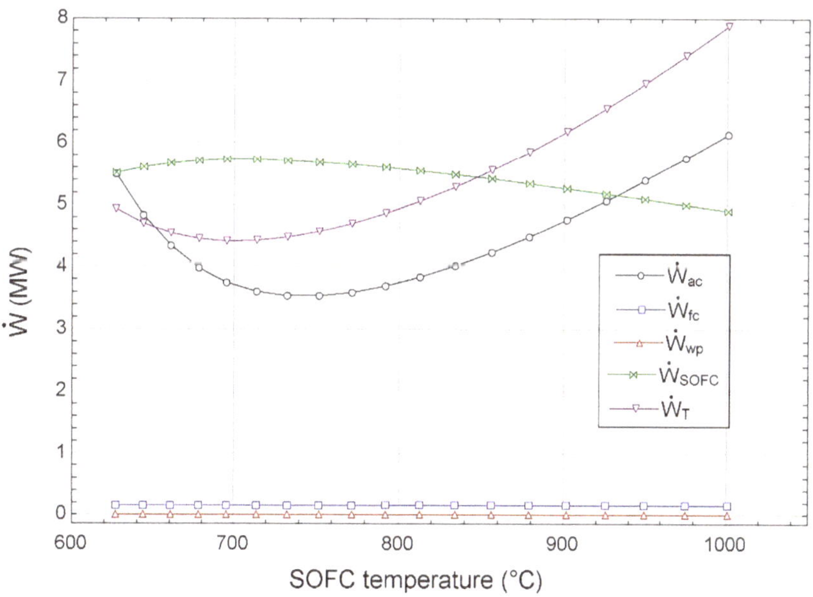

Fig. 7.30 Variations with SOFC temperature of electrical power productions and consumptions for components of the hybrid SOFC-battery-gas turbine system.

observed that the electrical power produced by the SOFC increases slightly as the SOFC temperature reaches 700°C, and then starts to decrease above this temperature. The reason is that the voltage (or potential) losses in the SOFC—including ohmic, activation, and concentration overpotentials—decline as the temperature rises to 700°C, and then rise as SOFC temperature further increases. In addition, the increase in the voltage losses with increasing SOFC temperature up to 700°C results in an increase in the heat loss in the SOFC in which the energy of SOFC products declines. This leads to a reduction in the electrical power production by the gas turbine as can be observed as the SOFC temperature rises to 700°C. At that temperature, the electrical power produced at the gas turbine reaches its minimum value.

Fig. 7.31 shows the variations in net electrical power produced by and energy efficiency of the hybrid system, with SOFC temperature. The net power production and the energy efficiency curves for the system exhibit very similar profile variations with SOFC temperature. Both parameters reach their maximum values at an SOFC temperature of approximately 775°C. So, it is suggested that the SOFC operates at this temperature. The net electrical power produced by and energy efficiency of the hybrid system increase respectively from 4.75 to 6.6 MW and 39.5%–46.2% as the SOFC temperature increases from 627°C to 775°C.

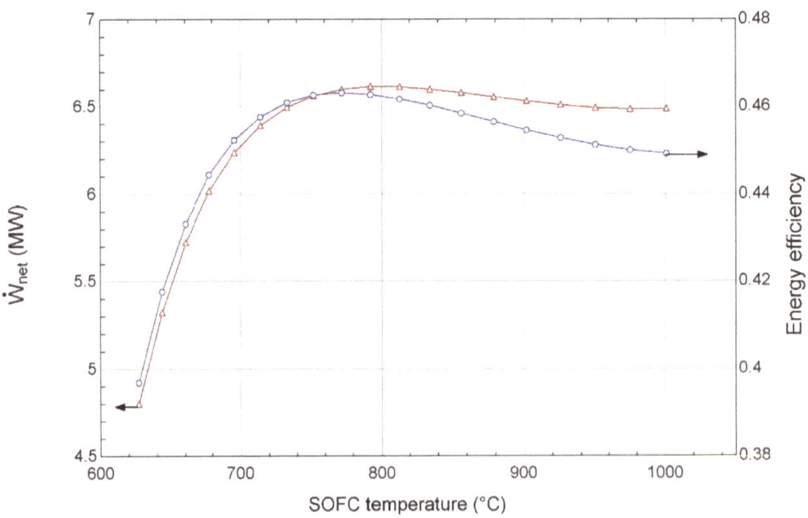

Fig. 7.31 Variations with SOFC temperature of net electrical power production and energy efficiency for the hybrid SOFC-battery-gas turbine system.

7.5.4 Closure and future research

In this case study, a novel design of a propulsion system for an electric aircraft is described in which the SOFC-battery is hybridized with the gas turbine. Lithium-ion batteries are used to supply electrical power during operating conditions having high power loads like climb and takeoff. The system is designed accounting for the cooling process required for the lithium-ion battery pack. To increase the lifetime and active operation of lithium-ion batteries in electric aircraft, battery thermal management systems are required to provide cooling for the battery cells. In the case study, water input for use in the SOFC is first pumped through the battery pack to remove heat generated in the battery cells; the water is then transferred to the heat exchangers and finally to the SOFC. The thermal analysis of the lithium-ion battery cell determines the temperature and heat generation rate within the battery cell. Also, the effect of C-rate on the temperature and heat generation rate is described. The average values for the temperature and heat generation rate are used in the simulation of hybrid-electric propulsion system. Through a parametric analysis of the system and its components, the effects are described of varying several operating conditions, such as SOFC temperature and pressure and current density, on the net electrical power produced by the system and its energy efficiency. Optimum values at which the maximum net power production and energy efficiency are achieved are identified.

Further research investigating on optimum battery energy storage capacity, airframe design and dynamics, use of pressurized SOFC, system packaging and aircraft subsystem are required for commercial implementation and to ensure safety and reliability of the aircraft. Also, designing light-weight battery thermal management system could improve achievable payload. Overall, the analysis of the proposed hybrid propulsion system for electric aircraft suggests that SOFC-battery-gas turbine hybrids may have a role in future commercial aircraft.

7.6 Closing remarks

In this chapter, the integration of battery-based technologies is described, and two case studies are presented that discuss novel types of battery-based integrated system technologies. The first case study examines the integration of a battery thermal management system with a fuel cell/battery electric vehicle. In the second case study, an electric aircraft propulsion system including a SOFC, a lithium-ion battery and a gas turbine is considered. In the first case study, the heat generation rate in the battery is determined under isothermal conditions, while, in the second case study, the heat generation in the battery is determined under a nonisothermal condition where the temperature of the battery is allowed to change. Comprehensive modeling, analyses, and assessments of each system are provided, highlighting insights into the development of new battery-based systems. The case studies provide the detailed electrochemical analysis equations implemented for battery and fuel cell subsystems to evaluate the overpotentials and irreversibilities of those subsystems. Further, the performances of the systems are assessed at varying operating conditions and system parameters through parametric studies. The electrical power output and the system performance are elucidated with recommendations for further improvements.

Study questions

7.1. An integrated battery-based energy system is shown in Fig. 7.32 which includes photovoltaic (PV) arrays, a water electrolysis unit, a PEM fuel cell, a battery pack, hydrogen and oxygen storage tanks, a converter and an inverter. The integrated system is used to supply the electrical power for a house. A part of excess electricity produced by the PV panels is stored as chemical energy in the form of hydrogen through

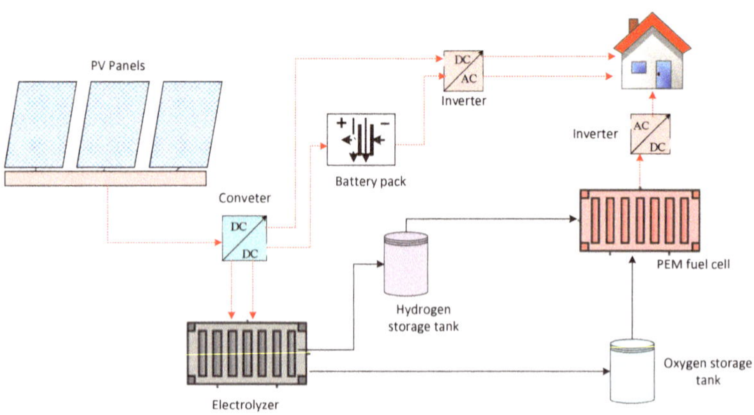

Fig. 7.32 Integrated PV-PEM fuel cell-battery system to supply electrical power for a house.

a water decomposition process in the form of electrolysis. Water is split into hydrogen and oxygen in the electrolysis unit by electricity. The other portion of the electricity generated by the PV cells is stored in battery cells so that it can be supplied during periods of poor solar radiation. The hydrogen and oxygen are combined in the PEM fuel cell for electrical power production. If the energy efficiency of an energy system is defined as the ratio of useful energy rate output to the total energy rate input, determine the energy efficiency for this integrated battery-based energy system. Make any reasonable assumptions you deem necessary.

7.2. A renewable energy-based electricity and freshwater production plant is shown in Fig. 7.33. The electrical power produced from the wind turbines is used by the reverse osmosis (RO) desalination unit for freshwater production. A water storage tank ensures a continuous supply of feed water. When the electricity produced from the wind turbines exceeds the electricity required by the RO plant, the excess electricity is stored in the battery and then recovered when needed by chemical-to-electricity conversion in the battery. A constant average value can be considered for the heat generation rate in the battery pack during charge and discharge. The seawater, which initially enters the pretreatment unit, is used to cool the battery pack before entering the RO desalination unit. If the overall energy efficiency of the energy system is defined as the ratio of useful outputs (freshwater production rate and

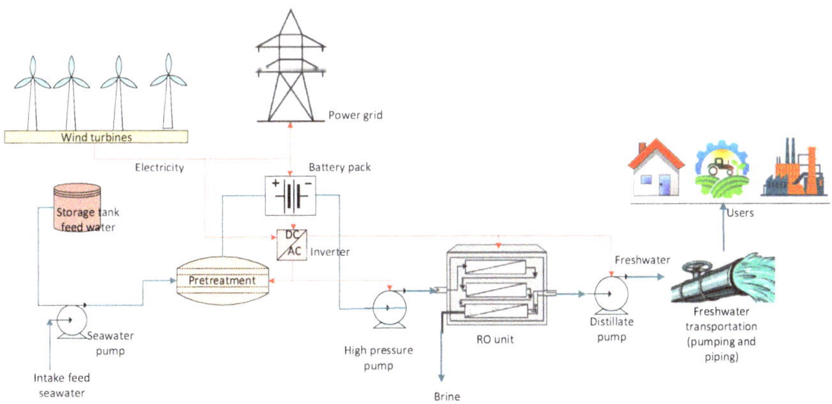

Fig. 7.33 Freshwater supply from a reverse osmosis (RO) seawater desalination plant using wind energy as the primary energy source. The seawater, which initially enters the pretreatment unit, is used to cool the battery pack before entering the RO desalination unit.

electrical power) to inputs (intake seawater and wind power), calculate the overall energy efficiency of this renewable energy-based system. Make any reasonable assumptions you deem necessary.

7.3. Redo Study Question 7.1 on an exergy basis, rather than on an energy basis.

7.4. Redo Study Question 7.2 on an exergy basis, rather than on an energy basis.

References

Al-Zareer, M., Dincer, I., Rosen, M.A., 2019. Comparative assessment of new liquid-to-vapor type battery cooling systems. Energy 188, 116010.

Amphlett, J.C., Baumert, R.M., Mann, R.F., Peppley, B.A., Roberge, P.R., Harris, T.J., 1995. Performance modeling of the Ballard mark IV solid polymer electrolyte fuel cell. I. Mechanistic model development. J. Electrochem. Soc. 142 (1), 1–8.

Chen, K., Hou, J., Song, M., Wang, S., Wu, W., Zhang, Y., 2021. Design of battery thermal management system based on phase change material and heat pipe. Appl. Therm. Eng. 188, 116665.

Cho, J., Jeong, S., Kim, Y., 2015. Commercial and research battery technologies for electrical energy storage applications. Prog. Energy Combust. Sci. 48, 84–101.

Collins, J.M., McLarty, D., 2020. All-electric commercial aviation with solid oxide fuel cell-gas turbine-battery hybrids. Appl. Energy 265, 114787.

Corrêa, J.M., Farret, F.A., Canha, L.N., Simoes, M.G., 2004. An electrochemical-based fuel-cell model suitable for electrical engineering automation approach. IEEE Trans. Ind. Electron. 51 (5), 1103–1112.

Doyle, M., Newman, J., Gozdz, A.S., Schmutz, C.N., Tarascon, J.M., 1996. Comparison of modeling predictions with experimental data from plastic lithium ion cells. J. Electrochem. Soc. 143 (6), 1890–1903.

Doyle, M., Fuller, T.F., Newman, J., 1993. Modeling of galvanostatic charge and discharge of the lithium/polymer/insertion cell. J. Electrochem. Soc. 140 (6), 1526–1533.

E-Bio fuel-cell, 2017. Available from: https://www.nissan-global.com/EN/INNOVATION/TECHNOLOGY/ARCHIVE/E_BIO_FUEL_CELL/. (Accessed 25 May 2022).

Farsi, A., Rosen, M.A., 2022a. Comparison of thermodynamic performances in three geothermal power plants using flash steam. ASME Open J. Eng. 1, 011016.

Farsi, A., Rosen, M.A., 2022b. PEM fuel cell-assisted lithium ion battery electric vehicle integrated with an air-based thermal management system. Int. J. Hydrog. Energy 47 (84), 35810–35824.

Farsi, A., Rosen, M.A., 2023. Performance analysis of a hybrid aircraft propulsion system using solid oxide fuel cell, lithium ion battery and gas turbine. Appl. Energy 329, 120280.

Fernandes, M.D., Andrade, S.D.P., Bistritzki, V.N., Fonseca, R.M., Zacarias, L.G., Gonçalves, H.N.C., de Castro, A.F., Domingues, R.Z., Matencio, T., 2018. SOFC-APU systems for aircraft: a review. Int. J. Hydrog. Energy 43 (33), 16311–16333.

Fuller, T.F., Doyle, M., Newman, J., 1994. Simulation and optimization of the dual lithium ion insertion cell. J. Electrochem. Soc. 141 (1), 1–10.

Gurau, V., Barbir, F., Liu, H., 2000. An analytical solution of a half-cell model for PEM fuel cells. J. Electrochem. Soc. 147 (7), 2468.

IEA, 2021. Energy Storage. IEA, Paris. Available from: https://www.iea.org/reports/energy-storage. (Accessed 25 May 2022).

IEA, 2022. Transport sector CO_2 emissions by mode in the Sustainable Development Scenario, 2000–2030. IEA, Paris. Available from: https://www.iea.org/data-and-statistics/charts/transport-sector-co2-emissions-by-mode-in-the-sustainable-development-scenario-2000-2030. (Accessed 25 May 2022).

Hoang, A.T., 2018. A review on fuels used for marine diesel engines. J. Mech. Eng. Res. Dev. 41 (4), 22–23.

Horizon, 2020. Smart, green and integrated transport. Work Programme 2018–2020. Available from: https://ec.europa.eu/research/participants/data/ref/h2020/wp/2018-2020/main/h2020-wp1820-transport_en.pdf. (Accessed 25 May 2022).

Jansen, R.H., Bowman, C.L., Clarke, S., Avanesian, D., Dempsey, P.J., Dyson, R.W., 2019. NASA electrified aircraft propulsion efforts. Aircr. Eng. Aerosp. Technol. 3, 1–13.

Ji, Z., Rokni, M.M., Qin, J., Zhang, S., Dong, P., 2020. Energy and configuration management strategy for battery/fuel cell/jet engine hybrid propulsion and power systems on aircraft. Energy Convers. Manag. 225, 113393.

Kalaf, O., Solyali, D., Asmael, M., Zeeshan, Q., Safaei, B., Askir, A., 2021. Experimental and simulation study of liquid coolant battery thermal management system for electric vehicles: a review. Int. J. Energy Res. 45 (5), 6495–6517.

Kohout, L., Schmitz, P., 2003. Fuel cell propulsion systems for an all-electric personal air vehicle. In: AIAA International Air and Space Symposium and Exposition: The Next 100 Years, June, Ohio, p. 2867.

Kousoulidou, M., Lonza, L., 2016. Biofuels in aviation: fuel demand and CO_2 emissions evolution in Europe toward 2030. Transp. Res. Part D: Transp. Environ. 46, 166–181.

Lin, J., Liu, X., Li, S., Zhang, C., Yang, S., 2021. A review on recent progress, challenges and perspective of battery thermal management system. Int. J. Heat Mass Transf. 167, 120834.

Mann, R.F., Amphlett, J.C., Hooper, M.A., Jensen, H.M., Peppley, B.A., Roberge, P.R., 2000. Development and application of a generalised steady-state electrochemical model for a PEM fuel cell. J. Power Sources 86 (1–2), 173–180.

Malik, M., Dincer, I., Rosen, M.A., 2016. Review on use of phase change materials in battery thermal management for electric and hybrid electric vehicles. Int. J. Energy Res. 40 (8), 1011–1031.

Oreizi, D., 2020. Four Types of Electric Vehicles, Charged Future. Available from: https://www.chargedfuture.com/four-types-of-electric-vehicles/. (Accessed 25 May 2022).

Ranjbar, F., Chitsaz, A., Mahmoudi, S.M.S., Khalilarya, S., Rosen, M.A., 2014. Energy and exergy assessments of a novel trigeneration system based on a solid oxide fuel cell. Energy Convers. Manag. 87, 318–327.

Rao, Z., Qian, Z., Kuang, Y., Li, Y., 2017. Thermal performance of liquid cooling based thermal management system for cylindrical lithium-ion battery module with variable contact surface. Appl. Therm. Eng. 123, 1514–1522.

Rosen, M., Farsi, A., 2022. Sustainable Energy Technologies for Seawater Desalination. Elsevier/Academic Press, United Kingdom.

Seyam, S., Dincer, I., Agelin-Chaab, M., 2021. Novel hybrid aircraft propulsion systems using hydrogen, methane, methanol, ethanol and dimethyl ether as alternative fuels. Energy Convers. Manag. 238, 114172.

Thampan, T., Malhotra, S., Zhang, J., Datta, R., 2001. PEM fuel cell as a membrane reactor. Catal. Today 67 (1–3), 15–32.

Toyota, 2020. Toyota Europe Newsroom. Available from: https://newsroom.toyota.eu/introducing-the-all-new-toyota-mirai/. (Accessed 25 May 2022).

Wang, F.C., Fang, W.H., 2017. The development of a PEMFC hybrid power electric vehicle with automatic sodium borohydride hydrogen generation. Int. J. Hydrog. Energy 42 (15), 10376–10389.

Wang, Q., Rao, Z., Huo, Y., Wang, S., 2016. Thermal performance of phase change material/oscillating heat pipe-based battery thermal management system. Int. J. Therm. Sci. 102, 9–16.
Zero emissions, 2020. Available from: https://www.maritime-executive.com/article/number-of-zero-emission-shipping-projects-nearly-doubled-in-past-year. (Accessed 25 May 2022).
Zhang, F., Lin, A., Wang, P., Liu, P., 2021. Optimization design of a parallel air-cooled battery thermal management system with spoilers. Appl. Therm. Eng. 182, 116062.

CHAPTER 8

Closing remarks and future directions for batteries and their thermal management

Chapter objectives

- Present an overview of the book.
- Describe and discuss the challenges and future directions for sustainable battery technologies.
- Provide general future considerations for battery technologies.

8.1 Introduction

In this chapter, an overview of the book is provided. Then challenges and future directions for sustainable battery technologies are described. Finally, some general future considerations for battery technologies are provided.

8.2 Overview of book

In this book, rechargeable battery technology is described in depth considering aspects of operation, analysis, performance, and applications, including relevant aging mechanisms and thermal management systems. The underlying fundamentals are detailed first to provide an understanding of the development and operation of rechargeable battery systems. Modeling and analysis of battery technologies are explained in detail, including coverage of overpotentials in battery cells and the use of thermal-electrochemical coupled modeling to determine temperatures and temperature distributions for battery cells and battery packs. Battery thermal behavior and its connection to aging, heat generation, thermal management and thermal failure are also described. Recent developments and advances in various types of battery thermal management systems are identified and discussed.

Furthermore, the main process steps for design of a battery system are described, including battery management system design, as well as electrical,

Battery Technology
https://doi.org/10.1016/B978-0-443-18862-6.00004-5

mechanical, and thermal design. This gives insights into important facets of these design areas such as the layout of the battery management system and its interference with the other subsystems, arrangements of battery cells in the pack, stress-strain analysis and vibrations in the structure of the battery pack, and the selection of a suitable battery thermal management system. These design areas of in many circumstances dependent on the given application. Finally, the integration of battery-based technologies in selected applications is presented, and their performances are analyzed and discussed. In addition, assessments are shown to determine how their performances are affected by varying operating conditions.

Rechargeable battery technologies (especially lithium ion batteries) are increasingly used in diverse stationary and moving applications, and these are covered in this book, as are the significant enhancements in their safety, performance and stability that have been achieved. Sustainability challenges associated with battery systems are becoming increasingly important and recognized, and these are examined here. These challenges include ensuring the availability of raw materials and their processing costs, the economics of and waste generation related to battery fabrication, and end-of-life battery and component management.

In the next section of this chapter, challenges and future directions for sustainable battery technologies are described and discussed.

8.3 Challenges and future directions for sustainable battery technologies

Many efforts are ongoing to address concerns regarding batteries. These efforts include modified and new chemistries for batteries that enhance specific energy. However, much work remains to be done. This includes efforts to assess and reduce capacity fade, self-discharge, and safety issues during the operation for modified or new batteries. For instance, results from studies on surface coatings and measures to increase the porosity of electrodes have been found to be promising for some types of batteries, and further research may lead to incorporation of these developments in future battery designs (Chaudhary et al., 2021; Somo et al., 2021; Ali et al., 2020; Kuang et al., 2019).

The abundance and availability of the elements used in the structure of a battery are two of the determining criteria for its cost and sustainability. Selecting widely available, cost effective and environmentally benign materials often is beneficial, but it does not guarantee that the final electrical energy storage system will be environmentally friendly and inexpensive. A full assessment must account for the fact that the raw materials must be extracted, processed and synthesized, manufactured and, where applicable, recycled, and these processes can have detrimental effects on the environment.

The total *environmental impact* (EI) of a battery cell can be expressed as follows:

$$\begin{aligned} &\text{Total environmental impact of a battery cell} \\ &= \sum_{\text{Components}} \text{EI of extraction of raw materials} \\ &+ \sum_{\text{Components}} \text{EI of synthesis processes} \\ &+ \sum_{\text{Components}} \text{EI of implementation in the system} \\ &+ \sum\nolimits_{\text{Components}} \text{EI of recycling} \end{aligned} \tag{8.1}$$

Much effort is needed on reducing or minimizing the total environmental impact of a battery cell to increase the sustainability of battery technologies. Elements selected for cells need to be cost-effective and environmentally benign, and new technologies applying emerging chemistries with reduced or minimal impact on the environment require development. To help achieve these aims, environmentally benign processes and abundant materials need to be incorporated in the design of new battery cells, and advanced environmental tools like life cycle analysis need to be employed.

For reducing the total environmental impact of battery systems, it is helpful if the chemical composition of the cell is selected so as to yield a higher energy density. For the electrolyte, for example, there is a tendency to employ a nonaqueous electrolyte (as is done in lithium ion battery cells) instead of an aqueous electrolyte (as in lead acid and nickel-based battery cells), since higher operating voltages can then be obtained. The versatility of lithium ion batteries in terms of material used for the anode and the cathode is one of the greatest advantages of these batteries compared to lead-acid and nickel-cadmium batteries. These advantages provide more possibilities for new designs of lithium ion cells that achieve high energy densities at low cost. Some recent research on the cathodes of lithium ion batteries has focused on designing new polyanionic compounds and improving layered oxides (Assat and Tarascon, 2018; Tian et al., 2020; Hua et al., 2020; Larcher and Tarascon, 2015). Note, however, that although the electrolyte of a lithium ion battery is nontoxic, the use of lithium-based electrolytes may contain fluorine-based anions, which are reactive and hazardous. So, research is needed to eliminate the use of fluorine-based salts.

As stated before, the availability of raw materials to meet long-term demand in one of the main concerns for the sustainability of battery technology. There are several socioeconomic and environmental sustainability issues associated with mining processes of raw materials used in fabrication of battery cells (e.g., lithium and cobalt). In addition, significant processing steps are needed to extract pure metals from ore which raises geopolitical concerns for the supply chain that can notably affect the prices of raw materials. The development and use of new relatively abundant materials used in the construction of battery cells can mitigate (but likely not eliminate) the supply chain challenges. For example, the use of nickel in the positive electrode material of a lithium ion battery instead of cobalt can diminish the dependency on cobalt resources, but it probably raises pressures on supplies of nickel.

Additionally, the replacement of materials used in battery cell components should be aligned with circular economy criteria for recycling and reusing battery cells. Such factors, which are a key part of supply chain, are important, especially with recent increases in applications of battery-based systems (like electric/hybrid electric vehicles). Certainly, waste management considerations are very important.

The fabrication of *composite electrodes* through combinations of materials may improve battery performance, by providing active materials, conductive additives, etc. to the anode and cathode. The selection of proper materials and their compositions in electrodes requires research and experimentation. Composites need to be formulated that, among other traits:

(i) have ingredients that are chemically compatible,
(ii) have stable compositions at various operating conditions,
(iii) yield good or optimum porosity, and
(iv) do not compromise the properties of electrode (e.g., ionic conductivity).

Low-temperature synthesis processes of compounds are preferred to high-temperature processes for the fabrication of battery systems, as they require less thermal energy and have lower impacts on the battery cell. Low-temperature synthesis methods include hydrothermal synthesis, microwave-assisted processes, sonochemistry, and ionothermal synthesis. These are described in detail elsewhere (Manjunatha et al., 2021; Zhu et al., 2021; Sud and Kaur, 2021; Recham et al., 2010; Tarascon et al., 2010; Recham et al., 2008).

The *recycling* of batteries is a key process if battery technology is to be sustainable and less intrusive environmentally. Unfortunately, this topic often receives little attention in assessments of energy related systems. The recycling of batteries can be based on the removal of intact electrode material and its use in the fabrication of new cells, or on constituent material recovery through hydrometallurgical and pyrometallurgical approaches. These approaches involve treatments of metals in high temperature furnaces and aqueous solvents (Harper et al., 2019; Newton et al., 2021). Descriptions and comparisons of the main recycling processes for battery cells have been presented (Beaudet et al., 2020).

Efficient and effective recycling methods can lead to the recovery of valuable materials used in the structure of the battery and reduce or minimize the environmental and ecological impacts associated with the disposal of batteries. Due to the chemical complexity of battery cells, recycling processes are expected to incorporate multistep treatments. To ensure recycling processes are environmentally friendly and economic as well as beneficial, in as reasonable a manner as possible, they should be assessed using advanced tools like *life cycle assessment* (LCA).

LCA, in general, is a standardized method used for evaluating and assessing the environmental impacts of a system, product or process over all stages of its life cycle. For example, for the manufacture of a battery system, environmental impacts are evaluated for the following steps: extraction of raw materials, production processes, transportation, installation, operation, maintenance, disposal and any other relevant activities. Fig. 8.1 illustrates the LCA concept and methodology for an electric vehicle. LCA is important for assessing and understanding battery technology. The ultimate goal of LCA for battery systems is to reduce or minimize the environmental impacts associated with the preparation, fabrication, use and disposal of batteries. Numerous recent studies have focused on LCAs of various types of batteries (Porzio and Scown, 2021; Sun et al., 2020; Le Varlet et al., 2020; Mohr et al., 2020).

8.4 General future considerations for battery technologies

Batteries constitute a widely applicable and very beneficial technology. On a practical basis, they provide necessary energy storage services today in various ways, while on a more contemplative level they provide the potential for numerous future beneficial applications. In this book, we focus on the former, although we at times delve into the latter.

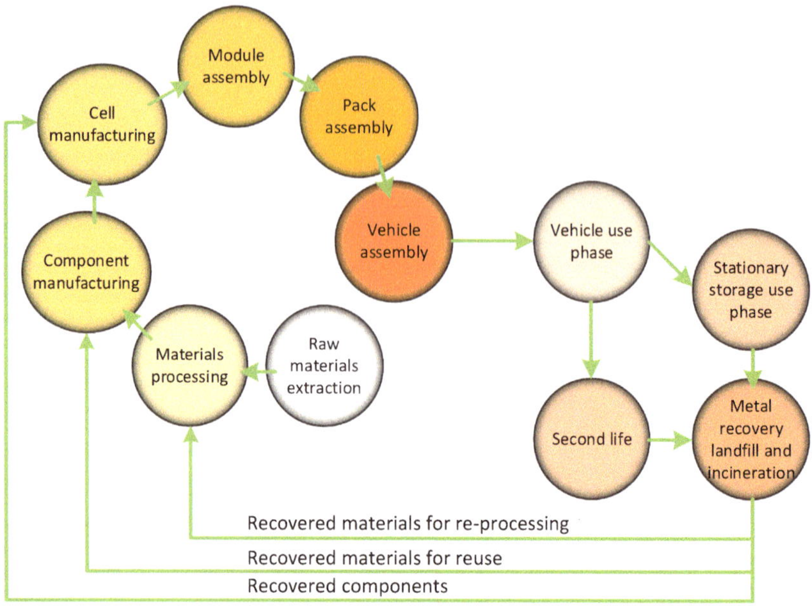

Fig. 8.1 Life cycle assessment concept and methodology applied to an electric vehicle.

The benefits provided by batteries and their future potential have attracted to the field in the past numerous researchers, who have contributed immeasurably to the development and application of battery technology. The attributes of batteries will, we are confident, continue to attract researchers and practitioners well into the future, and batteries will continue to evolve and find new applications.

In short, this book has sought to provide the reader with a broad working familiarity and understanding of batteries and their applications, by:

- Identifying and explaining battery technology and its use in improving energy management and efficiency, and designing better processes and systems.
- Discussing the broader aspects of batteries, in areas such as economics and environment as well as sustainability.
- Illustrating applications of batteries in various systems in hopes of explaining how to apply battery technology beneficially. The applications presented are by no means comprehensive, as many more applications exist, but are selected to provide a representative cross-section and to cover various important fields.

- Showing the many ways in which battery technology can assist efforts to achieve, or at least to shift toward, sustainability and sustainable development, in terms of such factors as use of sustainable resources and processes, increased efficiency, reduced environmental impact, improved economics, and others.

We feel that the benefits provided by batteries for industrial and other processes and systems are important in many ways. It is, therefore, worth developing an appreciation and knowledge of batteries, methods associated with them and their applications.

Much has been done to help increase the use of batteries over the last few decades, including the establishment of journals and international conferences, and courses and workshops on battery technology and its application. These activities complement well the impressive developments occurring in almost every area of science and engineering, from electrical and mechanical engineering to physics and from chemistry and chemical engineering to economics, and should ensure that batteries are utilized for enhancing advances in engineering and particularly energy.

We cannot help but contemplate about what the future may hold in store for batteries. Some expectations follow on how battery technology may be utilized, or how we hope it will be utilized, in the future so as to reap the greatest benefits:

- Batteries are likely to become increasingly utilized, across a diverse array of fields and throughout the developing and developed world, to increase use of renewable energy, enhance transportation, raise efficiency, reduce wastes and improve processes and systems.
- Battery assessments will become more integrated with environmental and ecological assessments, through tools like industrial ecology, LCA and others, and applied more widely.
- Battery technology will be used to foster sustainability and contribute to making development more sustainable.
- Battery technology will become more broadly covered in educational programs, to facilitate the development and adoption of enhanced applications, and to engender a broader understanding and appreciation of battery technology and its benefits by the public, by decision and policy makers, by leaders in industry and government.

If these expectations come to pass, even in part, numerous benefits will accrue to humanity and society through much better use and management of energy in particular, and improved processes, technologies and systems in general.

8.5 Closing remarks

In this chapter, an overview of the book is provided. Then challenges and future directions for sustainable battery technologies are described. Finally, some general future considerations for battery technologies are provided. For sustainable development of battery technologies in a wide range of applications further research is necessary in areas such as the availability of raw materials and their processing costs, battery economics, waste generation related to battery fabrication, and end-of-life battery and component management. Appropriate end-of-life battery disposal is a public health and safety issue. Recovering materials from spent battery cells and repurposing of end-of-first life batteries used in electric vehicles in a range of stationary energy storage applications are becoming increasingly important challenges for the future. Mining raw materials, materials processing, and manufacturing and recycling of batteries need to be considered in concert to achieve an economically viable and sustainable battery industry, and to allow batteries to deliver the multitude of benefits they can provide to society. These benefits include expanded use of intermittent renewable energy resources, enhanced management of energy systems, and increased adoption of electric vehicles, among others.

Study questions

8.1. What are two determining criteria for the cost and sustainability in the structure of a battery?

8.2. How can the total environmental impact of battery systems reduced? Provide an example to support your answer.

8.3. List and describe the main considerations in the fabrication of composite electrodes for a battery cell.

8.4. Why is recycling of battery cells important for their sustainability and for reducing the environmental impact of battery systems?

8.5. Describe the life cycle assessment approach. Explain how this approach can ensure recycling processes for battery systems are environmentally benign and beneficial?

8.6. Describe several ways in which battery technology may be utilized in the future to reap the greatest benefits.

References

Ali, E., Kwon, H., Choi, J., Lee, J., Kim, J., Park, H., 2020. A numerical study of electrode thickness and porosity effects in all vanadium redox flow batteries. J. Energy Storage 28, 101208.

Assat, G., Tarascon, J.M., 2018. Fundamental understanding and practical challenges of anionic redox activity in Li-ion batteries. Nat. Energy 3 (5), 373–386.

Beaudet, A., Larouche, F., Amouzegar, K., Bouchard, P., Zaghib, K., 2020. Key challenges and opportunities for recycling electric vehicle battery materials. Sustainability 12, 5837.

Chaudhary, M., Tyagi, S., Gupta, R.K., Singh, B.P., Singhal, R., 2021. Surface modification of cathode materials for energy storage devices: a review. Surf. Coat. Technol. 412, 127009.

Harper, G., Sommerville, R., Kendrick, E., Driscoll, L., Slater, P., Stolkin, R., Walton, A., Christensen, P., Heidrich, O., Lambert, S., Abbott, A., Ryder, K., Gaines, L., Anderson, P., 2019. Recycling lithium-ion batteries from electric vehicles. Nature 575 (7781), 75–86.

Hua, W., Schwarz, B., Azmi, R., Müller, M., Darma, M.S.D., Knapp, M., Senyshyn, A., Heere, M., Missyul, A., Simonelli, L., Binder, J.R., 2020. Lithium-ion (de) intercalation mechanism in core-shell layered Li (Ni, Co, Mn) O_2 cathode materials. Nano Energy 78, 105231.

Kuang, Y., Chen, C., Kirsch, D., Hu, L., 2019. Thick electrode batteries: principles, opportunities, and challenges. Adv. Energy Mater. 9 (33), 1901457.

Larcher, D., Tarascon, J.M., 2015. Towards greener and more sustainable batteries for electrical energy storage. Nat. Chem. 7 (1), 19–29.

Le Varlet, T., Schmidt, O., Gambhir, A., Few, S., Staffell, I., 2020. Comparative life cycle assessment of lithium-ion battery chemistries for residential storage. J. Energy Storage 28, 101230.

Manjunatha, C., Ashoka, S., Hari Krishna, R.H., 2021. Microwave-assisted green synthesis of inorganic nanomaterials. In: Inamuddin, Buddola, R., Ahmed, M.I., Asiri, A.M. (Eds.), Green Sustainable Process for Chemical and Environmental Engineering and Science: Green Inorganic Synthesis, Chapter 1. Elsevier, Amsterdam, pp. 1–39.

Mohr, M., Peters, J.F., Baumann, M., Weil, M., 2020. Toward a cell-chemistry specific life cycle assessment of lithium-ion battery recycling processes. J. Ind. Ecol. 24 (6), 1310–1322.

Newton, G.N., Johnson, L.R., Walsh, D.A., Hwang, B.J., Han, H., 2021. Sustainability of battery technologies: today and tomorrow. ACS Sustain. Chem. Eng. 9 (19), 6507–6509.

Porzio, J., Scown, C.D., 2021. Life-cycle assessment considerations for batteries and battery materials. Adv. Energy Mater. 11 (33), 2100771.

Recham, N., Armand, M., Laffont, L., Tarascon, J.M., 2008. Eco-efficient synthesis of $LiFePO_4$ with different morphologies for Li-ion batteries. Electrochem. Solid State Lett. 12 (2), 39.

Recham, N., Armand, M., Tarascon, J.M., 2010. Novel low temperature approaches for the eco-efficient synthesis of electrode materials for secondary Li-ion batteries. C. R. Chim. 13 (1–2), 106–116.

Somo, T.R., Mabokela, T.E., Teffu, D.M., Sekgobela, T.K., Ramogayana, B., Hato, M.J., Modibane, K.D., 2021. A comparative review of metal oxide surface coatings on three families of cathode materials for lithium ion batteries. Coatings 11 (7), 744.

Sud, D., Kaur, G., 2021. A comprehensive review on synthetic approaches for metal-organic frameworks: from traditional solvothermal to greener protocols. Polyhedron 193, 114897.

Sun, X., Luo, X., Zhang, Z., Meng, F., Yang, J., 2020. Life cycle assessment of lithium nickel cobalt manganese oxide (NCM) batteries for electric passenger vehicles. J. Clean. Prod. 273, 123006.

Tarascon, J.M., Recham, N., Armand, M., Chotard, J.N., Barpanda, P., Walker, W., Dupont, L., 2010. Hunting for better Li-based electrode materials via low temperature inorganic synthesis. Chem. Mater. 22 (3), 724–739.

Tian, Y., Zeng, G., Rutt, A., Shi, T., Kim, H., Wang, J., Koettgen, J., Sun, Y., Ouyang, B., Chen, T., Lun, Z., 2020. Promises and challenges of next-generation "beyond Li-ion" batteries for electric vehicles and grid decarbonization. Chem. Rev. 121 (3), 1623–1669.

Zhu, L., Chen, Y., Xiao, Z., Yang, H., Kong, L., 2021. Hydrothermal synthesis of star-shaped Bi_5O_7Br catalysts with strong visible light catalytic performance. J. Mater. Res. 36 (3), 628–636.

Index

Note: Page numbers followed by *f* indicate figures and *t* indicate tables.

F

G

H

I

J

K

L

FSC
www.fsc.org
MIX
Papier | Fördert gute Waldnutzung
FSC® C083411

Kolibri 360 GmbH
Ferdinand-Jühlke-Straße 7
99095 Erfurt, Deutschland
produktsicherheit@kolibri360.de